Copyright © 2017 Jorge Puell & Paloma Lazaro
All rights reserved
ISBN-10: 1974101991
ISBN-13: 978-1974101993

Mining & Water Quality:

Regulations and Practice for the U.S.

A review of the states under the "prior participation" doctrine

JORGE PUELL

And

PALOMA LAZARO

2017

Cover Photograph:
The Lavender open pit mine, Bisbee, Arizona
Photograph by Matthew P. Kowal

Table of contents

CHAPTER 1 .... **5**

MINING AND WATER 5
   I. INTRODUCTION 5
   II. MINE OPERATIONS 12
      1. Underground Mines 12
      2. Surface Mines 15
      3. Mine Development 21
   III. PROCESSING PLANTS 27

CHAPTER 2 .... **34**

WATER QUALITY CONTROLS 34
   I. INTRODUCTION 34
      1. Physical characteristics 38
      2. Chemical characteristics 39
      3. Bacteriological characteristics 40
   II. WATER RIGHTS 40
   III. AGENCIES 42
      1. Federal agencies 42
      2. Environmental agencies 52
   IV. SURFACE WATER 56
      1. Introduction 56
      2. Regulations 57
         2.1. Safe Drinking Water Act (SDWA) 57
         2.2. Clean Water Act (CWA) 60
   V. GROUNDWATER 68
      1. Introduction 68
      2. Aquifers 69
      3. Regulations 72
         3.1. Ground Water Rule (GWR) 72
         3.2. Revised Total Coliform Rule (RTCR) 75
   VI. REMEDIAL ACTIONS 77

CHAPTER 3 .... **83**

REPRESENTATIVE STATES REGULATIONS 83

- I. ARIZONA 83
    - 1. Overview 83
    - 2. Definitions 84
    - 3. Water Quality Controls 85
        - 3.1. Aquifer Permit Required 85
        - 3.2. Arizona Pollutant Discharge Elimination System Program (AZPDES) 91
        - 3.3. Enforcement 92
- II. COLORADO 93
    - 1. Overview 93
    - 2. Definitions 93
    - 3. Water Quality Controls 95
        - 3.1. General Provisions 95
        - 3.2. Water Quality Control Commission 96
        - 3.3. Procedures 99
        - 3.4. Permit System 99
        - 3.5. Penalties 102
- III. NEVADA 105
    - 1. Overview 106
    - 2. Definitions 106
    - 3. Water Quality Controls 109
        - 3.1. Permit for Facilities 109
        - 3.2. Discharge Permit 116
        - 3.3. Enforcement 119
- IV. WYOMING 121
    - 1. Overview 122
    - 2. Definitions 123
    - 3. Water Quality Controls 125
        - 3.1. Permits for discharges to surface waters 125
        - 3.2. Underground water protected 129
        - 3.3. Groundwater pollution control permit 129
- References 131

PREFACE

The U.S. mining industry has been considered, for many years, an outstanding and undisputed reference in the world. Terms and conditions for the mining existence have experienced profound transformations in the last decades equally involving regulatory reforms on environmental protection, safety and advances in equipment technologies. The United States environmental regulatory system leads the world in adopting mechanisms to set forth rules governing mining activities and the responsible use of resources from where the European Union and several other countries are identified and used similarly. In fact, U.S. mining corporations have expanded their scope to manage mines worldwide, dominating the mining industry and usually operating with the corporation's own environmental standards, provide theirs are more stringent than of those countries.

Water, under these considerations, is often the key element to highlight from the resulting convergence between environmental sustainable policies and mining operations. Water management practices are being implemented by company's environmental departments created with the purpose of attaining the most beneficial use of water, minimizing the risk of water pollution and preventing the discharge of contaminated water into the environment. Water is monitored and treated to ensure that its quality meets the regulatory criteria prior to being released. Failing to do so would cause irreversible damages on water-depending ecosystems, and consequently the revocation of any existing permits and licenses, large fines and other serious penalties, including imprisonment as well as endangering the company's reputation.

This book reviews the critical mining engineering practices related to water, U.S. legislation governing their interaction, agencies and other parties involved in these issues. Priorities have been given to those mining states located in the western arid zones (Arizona, Colorado, Nevada, and Wyoming), whose water rights law falls under the prior appropriation doctrine. This document is intended to cover the basics of the coexistence between water and mining. Many of the points presented here are purely descriptive and should be understood in a broader context, which goes beyond the scope of this book, for which the reader who wants in-depth understanding is recommended to look for further confirmation and clarification.

Jorge Puell and Paloma Lazaro

CHAPTER 1

MINING AND WATER

I. INTRODUCTION

Water is a major component in mining and metallurgical processes to recover minerals and each of these processes or operations use to a greater or lesser extent volumes of water to contribute to their efficiency. In some cases, the proximity or not to water sources may be a decisive factor for the economic feasibility of a mining project, provided that investments for access and plans to mitigate potential hydrologic effects may be on the sensitive edge of profitability. Since most mine-sites move and treat a steady amount of mineral material for certain long periods, the use of water is managed by maintaining its balance through recovering (reuse) and capturing (withdrawal) from new water sources. In mining, recovering water from its processes is generally preferable cost-wise and includes all water recovered, rerouted, reclaimed or recycled that was previously used within the mine property and that comes mainly from tailing ponds, interceptor wells, effluents from plant operations, tailing thickeners, coal washing facilities and concentrator overflows. Unlike new water withdrawn which is officially documented, quantities of recovered water is not publicly known and only managed internally by mining companies, although it represents significantly more volume than new water added to the circuit process. According to the U.S. Geological Survey report on water use (USGS, 2010), water withdrawals due to mining activities are estimated about one percent of total withdrawals nationwide in the United States (approximately 5,320 Million gallons/day were withdrawn for mining in 2010). All mining water withdrawals were considered self-supplied being groundwater the main source (73%) of total water rather than surface water (27%).

Depending on mineral production, mining requires hundreds of thousands of gallons of water in its circulating flows. Recovered rates in the process have been increased due to economic objectives and regulatory pressure in the last decades, and efforts to reduce effluents discharge to the environment have resulted in materialization of investment in treatment plants that return the water resources in sufficient quantity and quality. However, water losses due to transportation, vaporization or impregnation make it inevitable to enter new water to the processing circuit which brings up concerns on streams pollution and depleting groundwater. Regardless of the mining method, whether it is surface or underground, the most suitable water source is groundwater for several reasons, especially two of them: 1) In an open pit mine the water table aligns its level and inflows to the pit bottom and walls as mining advances deeper down. As a consequence, filtered water is collected and pumped from the bottom to the outmost levels to be used conveniently. 2) A similar process of filtration and difference of pressure and gravity occurs in underground mines. Access to the orebody is conducted by ramps, shafts and tunnels from where excess groundwater sunk into the mine is pumped to the surface. However, the type of mining operation may or may not relate to water issues upon the watershed, although unique mine characteristics (e.g. location, processing methods or metal type) vary to a great degree on this aspect.

In addition to these two main mining methods, there are a number of other mining techniques that are not very common but should be mentioned since water plays an important role in them, which include in situ solution mining, in situ leaching, auger mining, underground coal gasification and coal bed methane production. For instance, in situ solution mining is a technique applicable to some sulfur and salt deposits which react in solution with water without requiring explosives for fragmentation: boreholes drilled to the strata are injected with cold or hot water.

Historically, demand for metals has been beyond doubt or discussion. In recent times metals are used in a wide variety of industries and their applications have not only reached high levels of technical complexity but are fundamental piece in the world economy. Nonetheless, rapid development of large-scale mining industries since the second half of the twentieth century has aroused concern by local and governmental regulations on protecting the environment from all source of contamination that can be generated from now on, and a series of policies for environmental liabilities.

Major metal commodities produced in the United States

Domestic production of metal commodities, including precious and non-precious, are essential to the U.S. economy. A diverse supply of and demand for metal commodities occur nationwide depending upon mining characteristics of each particular state. The following table lists major produced commodities in the U.S. (as primary or by-products) and their main producing states.

Table 1.1. U.S. Mine Production Values in 2016 (USGS, 2016)

Commodity	Producing states	US$ billion
Coal	*Coal deposits are found mainly in the coalbed deposits of the Appalachian Region (Eastern U.S.) and the Wyodak coalbeds (Western U.S.) representing a major industry in the country Four largest producer are Wyoming, West Virginia, Kentucky and Pennsylvania.*	$31.3
Cement	*Mainly produced domestically (90%), cement plants are spread in 34 states for production of portland cement and masonry cement. Texas, California, Missouri, Florida and Alabama count for half the U.S production of cement*	$10.7
Potash, Soda Ash and Boron	*Potash is produced in New Mexico and Utah (US$680 million) used by the fertilizer industry* *Soda Ash production comes from Wyoming, California and Colorado. Soda ash (US$1.7 Billion), used mainly to make Glass, chemicals, soap and detergents* *Boron is produced by two companies in southern California. Production data has not been shared in 2016*	$0.4 $1.8 ---
Iron ore	*Iron ore is found in the Mesabi Range of Lake Superior as magnetite (oxide of iron). Open pit mines in Michigan and Minnesota produced and shipped 98% of the U.S. usable iron ore*	$3.4

Gold	*Major producers in the U.S are Nevada (which counts for two-thirds of domestic production), Utah, California and Alaska*	$8.5
Copper	*In 2016, twenty-four mines produced copper in the U.S. Major producer states are Arizona, New Mexico, Utah, Montana and Michigan (accounting for 99% of total mine production). Domestic copper consumption is about 66%*	$6.8
Lead	*Eleven lead mines produced the total domestic production, 5 of them as a coproduct. Missouri, Alaska, Idaho, and Washington are the major producers.*	$0.7
Zinc	*U.S. zinc production is mainly in Alaska and Missouri*	$1.7

Mine Sequence

Mining inversions driven to extract minerals often follow a series of activities towards the attainment of a final product. A typical sequence began with a land use planning which involves mine investors, community and authorities to decide where mining may or may not be appropriate on the landscape. Governments have established particular areas to ensure the protection of critical ecosystems and if there is space available a claim-staking supersedes to perform subsequent exploration activities. Afterwards, if a promising deposit is found, mine investors will conduct economic and engineering feasibility studies. If it is decided to move forward, the project would be subject to environmental assessments which vary according to the characteristics of the project. After approvals and permitting have been issued it will follow a mine development, mine operation (the cyclical active action of mining), mine beneficiation, closure and reclamation. In and of themselves, building a mine is a high capital investment and long-term undertaking that requires a permanent process of consultation to engage all stakeholders to obtain their consent to these mining activities.

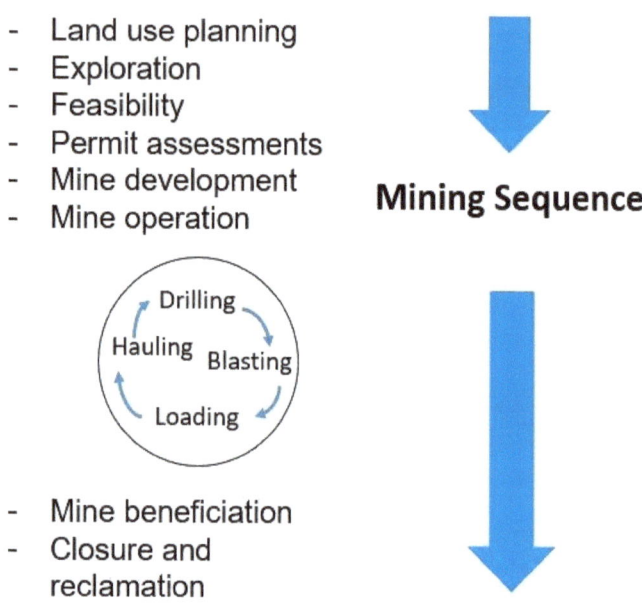

Figure 1.1. Typical mining sequence

Mining impacts on water

The main users of water are usually related to the ore processing (El Idrysy & Connelly, 2011). Nonetheless, environmental impact on water can be originated from various other sources found in the following mine processes:

- Mine operation: Surface or underground
- Leaching run-of-mine dumps

- Waste rock dumps and tailings dams
- Mineral processing plants
- Dewatering
- Untreated water discharge

The mining structure by itself represents a reconfiguration of the landscape and watershed. Identified risk events that can potentially disrupt and pollute downstream water are, among others, acid rock drainage, the risk of mine flooding, tailings dam failure, mine dewatering and diversion of hydrological paths.

Figure 1.2. Coal transportation at North Antelope Rochelle mine in Wyoming

II. MINE OPERATIONS

1. Underground Mines

Underground mines are referred to those excavations below the surface of the earth from which hard rock minerals are extracted by means of mine shafts, ramps and horizontal or vertical tunnels. Orebodies mined through underground methods generally contain high-grade percentage of metal which makes profitable the extraction of relatively narrow and stretchy veins. Different elements are considered to determine the most appropriate mining method for each mineralized body. Seam thickness, shape, overburden depth, orientation and type of the orebody define whether the selective mining method will extract rock from the stopes without filling the voids, leaving pillars or backfilling with waste rock, sand or tailings. Other factors that play an important role in the selection of an underground mining method are the capital financing, transportation and regulations.

Technological developments in the field have resulted in highly mechanized underground operations nowadays. Deepest mine shaft in the United States has been completed at Lucky Friday mine in Mullan, Idaho with a diameter of 5.5m and depth of 2,922 m below surface (Mining Magazine, 2016)

Underground Mining methods
Depending upon geological conditions, technical feasibility or effectual and economic restrictions a suitable underground method or a combination of them should be selected. For coal, most mines in the U.S. are conducted by underground methods, especially room and pillar or longwall.

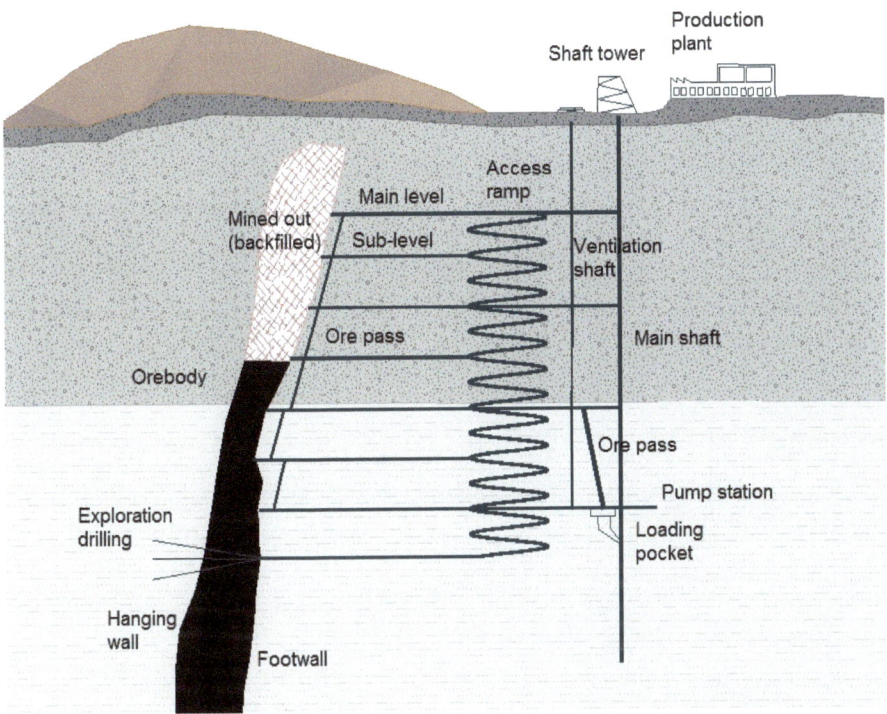

Figure 1.3. Schematic layout of an Underground mine

- Room and pillar is a method designed to provide roof support and prevent subsidence by leaving behind supporting pillars between hollowed rooms mined in a, generally, flat or lightly dipped coal seam. One of the major drawbacks of this method is the loss of unrecovered reserves left in the pillars. Room and pillar is the most traditionally used method for underground coal mines in the country.
- Longwall mining is the most common mechanized underground methods to extract coal from flat coalbeds. It uses large and relatively expensive equipment: continuous miner, conveyor systems, coal shearers and more infrastructure. The method consists of extracting large panels of rectangular blocks of coal, often larger than 1,000 feet wide and two miles long,

which are shear and spill into the conveyor to transport the coal to the surface and provide access to enter the room.

Subsidence

Mine subsidence occurs when voids that were left underground as a result of underground mining workings collapse and propagate to the surface, causing sinkholes, troughs and other related features that can cause structural damage and property damage on buildings, roads and utilities due to movements, cracking, shearing and land distortion in response to stress and settlement of the surface. Subsidence can occur over very old mines in an unpredictable manner (Lee and Abel, 1983) since no warning signal is possible from unmined pillars left in abandoned mines.

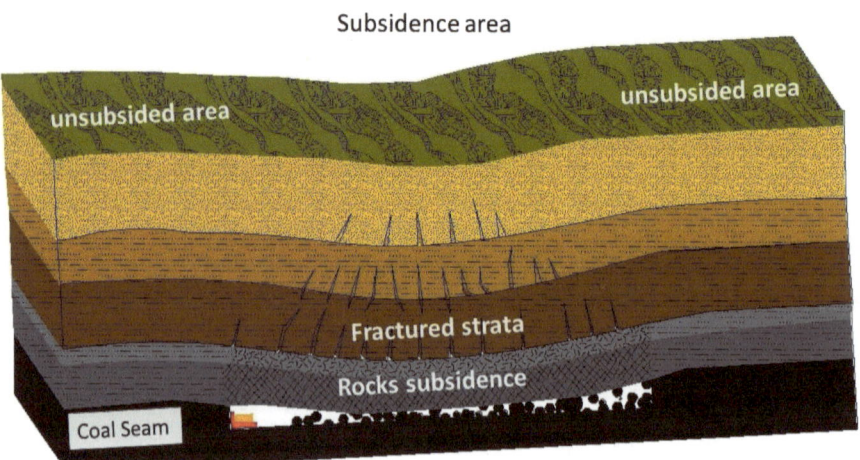

Figure 1.4. Surface deformation due to mine subsidence

Energy

The U.S. mining industry consumes about 1,250 Trillion Btu/year and is classified as an energy-intensive industry (U.S. Department of Energy, 2007). Mining activities identified as major energy consumers are blasting, drilling, ventilation, dewatering, crushing, grinding, transportation and separation processes. In underground mines, electricity is generally the main source of energy and requirements are significantly higher than surface mines on an energy consumption per ton basis, given the large amounts of tonnage moved in surface mines in contrast with the selective mineral extraction related to underground methods (U.S. Department of Energy, 2002). Underground mine operations consume great amounts of energy for ventilation, hosting and pumping power to transfer water from within the mine to a processing plant or a discharge location. On the other hand, the majority of surface mines energy consumption is on diesel fuel haul trucks.

2. Surface Mines

Surface mines, also known as open pit mines, are mines in which the orebody is mined from the surface of the earth with outdoors personnel and machinery exposed to the outside air. This method allows the mineralized deposit to be mined from stepped horizontal benches which generally vary between 9 and 30 m height and the removal of overburden material requires an ex-pit disposal site. A mining method so explained is suitable for orebodies emplaced near the surface that is for the ore and overburden relationship to be yet of economic interest. This mining form differs from the underground method in which the ore deposit is found deep below the surface and requires more selective mining through tunneling into the earth.

Depending upon the ore type, large-scale surface mines can be developed in situations where the stripping ratio of overburden to ore is as low as 1:10, so that waste moved from the pit would be sufficient to backfill the void. By contrast, shortage of waste to backfilling because of high stripping ratios would lead to abandoned surface mines to form lakes.

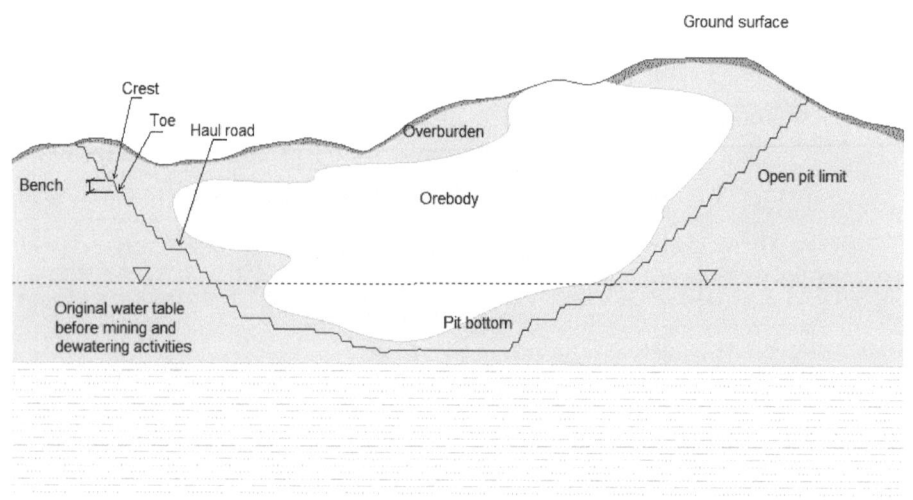

Figure 1.5. Typical cross section of a surface mine

Quarries

While the term open pit usually refers to metallic or non-metallic orebodies, a quarry is a type of open pit establishes to extract dimension stone, construction aggregate (sand and gravel), granite, limestone, or slate. Quarries have smaller and almost vertical benches.

Opencast mines

Opencast mining, also known as strip mining, means a surface mine method to work bituminous coal. It differs from the open pit in geological and operational respects. Coalbeds are emplaced in layers between the surface and internal overburdens, with typical strip ratios of overburden over coal beam of 20:1 which makes uneconomical to move out all the stripped overburden ("spoil") to external dump areas, but brings forth enough room to "cast" the spoil by backfilling the area right after stripping. As the coalbeds are rather long than scattered, the final pit left does not resemble an inverted cone as in a metallic open pit mine.

Figure 1.6. Coal seam mining by rope shovel and truck. North Antelope Rochelle mine in Wyoming

Mine rock dumps

A challenging fact about surface mining is that there is a significant amount of waste produced as the concentration of metal in the ore is susceptible to the economical cutoff grade. Consequently, the marginal and uneconomic material becomes run-of-mine or waste rock mine that is stored in a convenient dump place.

A mine rock dump is a massive structure made of run-of-mine or waste material piled in an external area of the mine or thrown in an already mined pit void (in pit dump). These areas ended up to be fairly larger than the pit itself, affected by an increase in volume after they are released from the earth (swelling factor) and subject to slope stability controls by the rock angle of repose.

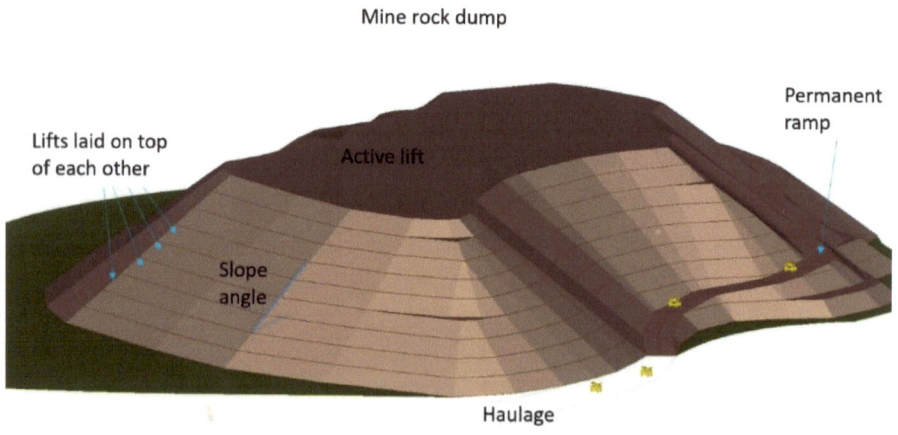

Figure 1.7. Typical configuration of a mine dump

In pit dumping

Large mining districts commonly work more than one open pit within their limits, making it possible to refill exhausted pits, thereby reducing the need to acquire an ex-pit land to stockpile mined material above the surface grade. In pit dumping has the advantage to reduce environmental impacts that have been already addressed during the initial phases of operation (Marcus, 1997), as well as existing hydrogeologic information from engineering and exploration feasibility studies and a reduced impact on groundwater levels by the fact that the material returns to its original position. Nonetheless, adverse construction and environmental conditions may offset the advantages by other existing related issues, which include:

- Burying potentially economic mineral resources under a future scenario of higher commodity prices
- Installation of impermeable liners to cover steep high walls if required
- Construction of ponds and collection impoundments in the pit bottom
- Stability concerns during the unloading activity, frequent sinking of the dump face due to significant pit height

Pit Lakes

Final open pit limits are of various sizes and depths and usually reach the water table at the bottom benches. Continuous pumping is required to keep the mine dry for operations. Post-mining reclamation activities, however, could allow the pit to fill naturally or artificially (Gammons et al., 2009) by flooding or through precipitation or groundwater infiltration (Castro and Moore, 2000). Thereby, a pit lake can be defined as a lake formed by flooding of an excavated open pit.

Figure 1.8. Pit lake at Anaconda Mine in Nevada
(Photo taken from Wikipedia.org)

Figure 1.9. Pit Lake formed after Berkeley mine' closure
(Satellite image taken from Google Earth Pro, 2017).

Figure 1.10. Pit Lakes at the former Gilt Edge open pit mine, South Dakota. Cleanup, environmental reviews and interim remedies are conducted by EPA (Satellite image taken from Google Earth Pro, 2017).

3. Mine Development

Mine development is the area responsible for providing support to the mine with respect to mine site preparation, construction of facilities, hold up on maintaining land-use/environmental permits, the creation of infrastructure for water, substations and power lines, fueling docks, mechanical shops, administrative buildings and other auxiliary services. These activities are managed through engineering projects of varying duration and given the diversity of specialties involved they are entrusted to third parties whether they are consulting firms or contractors.

Engineering basis for water management is developed at this stage of the mine cycle. The main mining facilities and water systems built to manage water uses are, among others:

Dewatering systems

Dewatering systems are unavoidable in almost all mining operations that reach lower levels than the water table. A hydrogeological study is required to select the most appropriate dewatering technique for the mine site, which includes:

- Mine pumping systems selection and projecting pumping long-term plans by estimating head requirements, quantities of water inflows, pump type, number and location of main and subsidiary pumps and the elaboration of contingency plans (Bridgwood et al., 1983). In open pits, pumps are usually equipped with floated platforms on the holding reservoir, in many cases the pit itself, which if associated with hydrological regime can encourage the formation of mine influenced water (Gusek & Figueroa, 2009). Suction lift and vertical distance are limited, hence deep mines need to submerge pumps and connect them in series on various levels of the mine, pushing up from lower to higher levels and out of the mine, assuming the risk that the entire system functionally is depending upon each one of the pumps (Morgan, 2017)

- Sumps designated to accumulate and temporarily store groundwater inflows in the excavation which are constructed of a non-earthen material (e.g. concrete, steel, plastic or any material that is impervious to the solution being collected) that provides structural support. Sumps are equipped with robust pumps capable of handling water containing significant suspended solids. Sumps may require extensive and costly monitoring as well as registration with the regulatory agencies.

- Intercepting wells are used to intercept lateral groundwater flow into the pit and to lower groundwater levels in advance of mining. Around waste dumps or tailing dams, intercepting wells provide containment of discharge pollutants to groundwater, tunnels, sewer, water trenches and other buildings. Slope stability can be improved by using this technique. In underground mines, they reduce roof pressures on temporary supports and sheeting (Sterrett, 2007). Monitoring wells and piezometers installed in the vicinity of the mine site prior to dewatering provides valuable data on hydrogeological conditions.

- Slope depressurization by using inclined or horizontal drains to provide permeable pathways to allow trapped or slowly draining groundwater behind pit slopes to bleed off into the pit

- Grouting and artificial ground freezing are used to seal off preferential groundwater pathways, or to reduce ground permeability in advance of shaft sinking or roadway development.

Most mines adopt a 'zero discharge policy' in which dewatering effluents are stored in reservoirs and may receive some form of treatment. Mine development provides resources to install and maintain all dewatering sumps, including pumps, piping, sump pits and backfill and other facilities for control, collection and disposal of groundwater.

Figure 1.11. Sumps at bottom benches of the Bingham Canyon mine, Utah (Google Earth Pro, 2017)

In surface mining, piping lays out along the pit wall and ramps although is buried at haul road crossings, while in underground mines the piping size and routing is considered in all sections of permanent and temporary access tunnels to water sources. High pressures at deep underground elevations create striving conditions such as pumping slurry and extending shaft space. Dewatering systems shall prevent loss of fines, boiling, softening of foundations strata and maintain excavation's stability of the bottom so that every part of the work can be performed in the dry. For this purpose, the dewatering system should take into account estimations of hydraulic parameters, mining and pumping rates and stormwater calculations to develop a dewatering model.

Non-storm water reservoirs

Water removed from mining areas needs a space for disposal if a natural and convenient water course is not found nearby. Depending on size and capacity, certain reservoirs may require special permits. Non-storm water reservoirs are surface or underground water impoundment designed for intermittent use where stormwater, if any, flows commingle with mine drainage or process wastewaters. Non-storm water facilities are installed at strategic locations and its operation and maintenance should be inspected frequently to remove excess debris that may impair proper functionality, that is, water levels should be pumped to the lowest possible levels for cleanup. Surface impoundments used in mining industrial circuit with no significant risk for surface water run-on or with containment capacities which far exceed the containment requirements for a 100-year, 24-hour storm event. Maintenance includes testing of water quality by adjusting pH of fluids to a pre-defined level.

Process solution reservoirs

Storage facilities where water is stored, reclaimed and returned to the circuit solutions, for dust control or reclamation (e.g. emergency ponds, raffinate, acid plants treatment, detention ponds). Most large-scale mines operate more than one of these facilities, which requires frequent monitoring and maintenance to detect and mitigate potential circumstances that may cause a discharge. Individual reservoirs should operate at a certain minimum or maximum levels based on site-specific operations.

If a stormwater run-on to the facility is detected, an alert should indicate that a maximum level hasn't been exceeded. These maximum levels are established to allow sufficient freeboard at each reservoir to contain a 100-year 24-hour storm runoff water volume.

Collection channels
Surface water and leach solution control ditches, berms, culverts or other structures will be monitored during scheduled maintenance inspections for erosion, sedimentation, breaches, clogging from debris or vegetation, or other degradation. Special attention will be paid to these components during and after the monsoon and heavy winter rainfall events.

Water Quality Sampling
Surface water must be sampled and controlled to maintain efficient mine operation and comply with regulations. A diversified and technically trained team conduct site assessments to provide a reliable representation of conditions and accurate data for decision making. Applicable sampling procedures and health and safety procedures practices are followed in order to measure field parameters on water quality, which include:

- Water alkalinity (pH) or the measure of the hydrogen ion concentration in water which should have values greater than 7.

- Conductivity (read out in µS/cm). "Pure" water conducts water very poorly, hence impurity in water increases its conductivity.

- Dissolved oxygen (or oxygen saturation) is a relative measure of the amount of gaseous oxygen dissolved in an aqueous solution. When dissolved oxygen levels drop below 5.0 mg/l, aquatic life is put under stress. Large fish kills can occur if oxygen levels remain below 1-2 mg/l for a few hours.

- Residual Chlorine is used to protect the public health by killing the microorganisms found in water which cause diseases. The chlorine residual test is used to determine the total amount of chlorine present as a residual (the amount of chlorine present after the demand has been satisfied). Because the residual determines how effective the disinfection process is, it is important to make sure that the residual remains within a specified range. Too little chlorine will not give adequate disinfection and too much can kill the aquatic life in the receiving waters.

III. PROCESSING PLANTS

Water used by mineral processing plants represents the highest consumption of water in relation to the mine-site total volumes, though their quantities vary to a significant degree by plant type and size operation. Water is used not only as the transportation medium of solids from different processes but it is also the fluid medium in which the mineral separation takes place. Mineral treatment at processing plants involves primary and secondary crushing, grinding following by flotation, classification and thickening and water flows internally and externally between these unit processes.

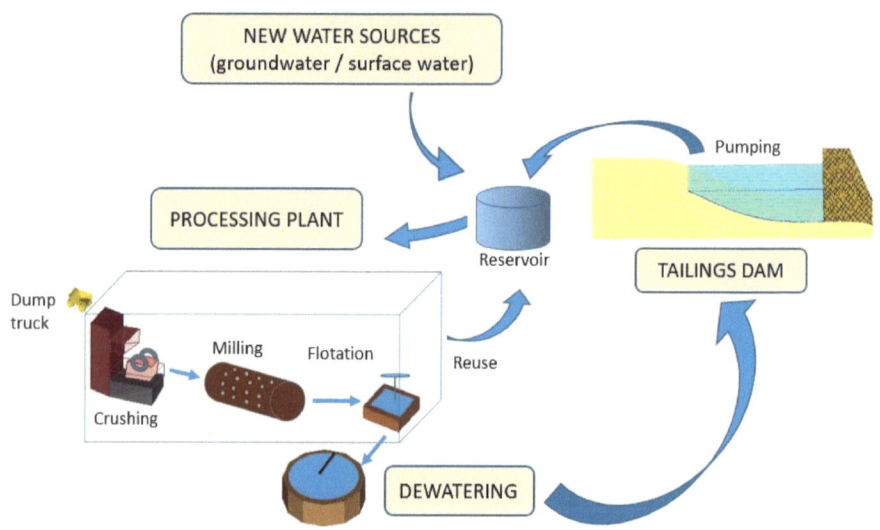

Figure 1.12. Closed water recirculation system in a mine processing setting

For instances, flotation operations operate at about 25 to 40% solids by weight. For a 40,000 tons per day concentrator that produces copper and molybdenum concentrates in New Mexico, about 105 m3/min of water is required, which is expensive to find and key to reuse. Diluting and dewatering mass flows in the closed circuits should be performed for optimum performance, and water balance is used to estimate water requirements and to determine circulating loads. Greater water consumption among these processes occurs in plants that employ wet separation methods (e.g. flotation, leaching, flocculation, gravitational, agglomeration and so forth), as well as in transportation of concentrates and tailings, evaporation and reservoir seepage.

Usually, ore receives pre-treatment before milling by means of adding water and reagents important for flotation. Flotation plants need an excess of water compared with the mineral volume where mineralized ore reacts in an alkaline solution reached by the reagents properties (from 7 to 10 or 11 pH in the flotation cells). Flotation final product is a concentrate with 20-40% metal content and a rate of 25% to 40% solids to obtain a higher recovery of the mineral. Water requirements during flotation can vary between 3 and 1.5 m3/tons of mineral.

Tailing dams

Tailings dams were originally introduced for water pollution control, however, they are now mostly used to prevent the released of silt (Younger, 2002) generated from the processing plants and to accumulate water to be reused in the circuit. In general, these dams are located downstream of mining operations (ANA, 2013), and are designed to retain the finest sediments. Tailings dams are built with residual soils taken from the mine itself or nearby places, and then properly moisture, compacted and prepared.

In regards to the tailings dam long-term stability some critical aspects are to be considered:

- Saturated tailings can generate high excess pore pressures if loaded or raised quickly.
- Tailings can be susceptible to internal erosion and piping
- If saturated and subject to high ground motion, tailings can be susceptible to seismically induced liquefaction
- Well prepared foundation conditions enhance overall stability performance
- Tailings can be characterized using conventional undrained soil mechanic principles.
- Tailings can be readily eroded by water (Fig 1.14)

Figure 1.13. Tailings impoundment dam at Fort Knox gold mine in Alaska (Google Earth Pro, 2017)

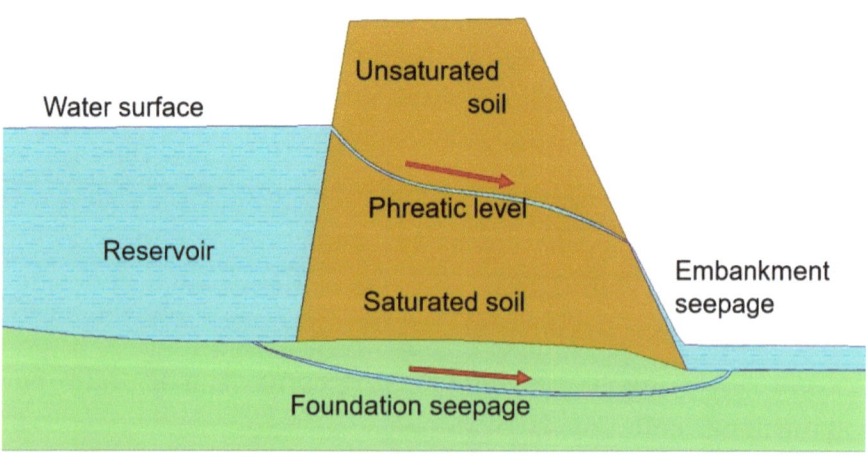

Figure 1.14. Seepage path through embankment and foundation

Water Balance

Developing a sound water balance model is fundamental for the water management team to make good decisions when facing challenges on water-related issues such as water shortages (especially in deserts zones), abundances during rainy seasons, frozen storages at extreme cold sites, monsoons and impacts product of changes in the mine operation itself (e.g. plant extensions, mining new phases, introducing new stockpiles and reclamation projects). Modeling the water balance allows to proactively identify opportunities for water management improvement.

The basic equation equals demand, supply and gain/loss (change in storage) to ensure that no artificial flows are created but maintained in the system:

Between the inflows and outflows is the process of the various mine elements (open pit or underground mine dewatering, processing plants, tanks, reservoirs, tailings dams, waste rock dumps or stockpiles, leach pads, collection ponds, mine town, mine operations services, treatment ponds and other infrastructure). The water balance should equal the relationship of inflows, outflows, and change in storage, where:

i) Inflows such a precipitation, runoff, dewatering from mine, supply wells from pumping stations, and

ii) Outflows such as evaporation, water discharges and seepage from tailings.

Preparing a water balance should consider:
- Multidisciplinary collaboration across mine departments
- Use of water data sources, automated instrumentation such as flow meter, historical climate variables, laboratory assays, facility's capacities, field surveys.
- Use of simulation software to predict water requirements for processing in 1-year, 5-years or life of mine periods. The model should be able to support deterministic and probabilistic simulations. Run sensitivities.
- Manage uncertain operation constraints such as complex routing of solution and pumping capacities
- Surface water drainages estimations
- Model calibration revisions periodically, addition of new data such as fresh water distribution and inflows/outflows from tanks. Prior to using the model, it must be calibrated by comparing projected and actual performance.
- Use of the water balance model as a primary decision-support tool for water issues.
- Use of global constants inputs which will not change such as the water density value of 8.34 lb./gal (at 40 ° F)
- Consider uses of water in the pit (dust control at crushers or haul roads), mine offices and unmetered flows for other operations based on estimates.
- Understanding active leach systems inflows/outflows for Leach cycle times and Non-dumping Leach cycle times.

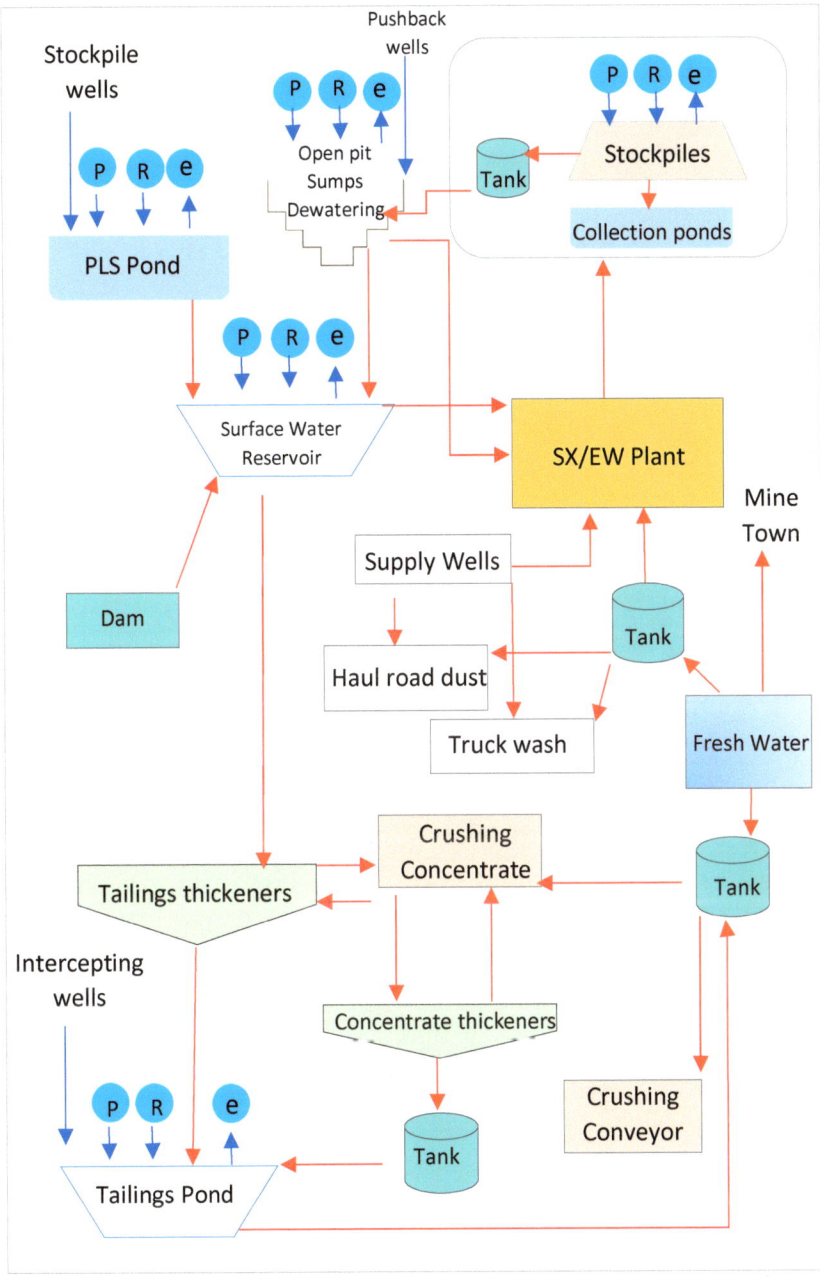

Figure 1.15. Typical Water Balance Model diagram for a copper open pit mine, where P= precipitation, R= runoff and e= evaporation

CHAPTER 2

WATER QUALITY CONTROLS

I. INTRODUCTION

A fundamental part in the feasibility of a mining project is obtaining governmental approval in the form of environmental permits or environmental impact assessments of many aspects of the mining plan, including those for design and construction of tailings dams, impoundment ponds, sewage treatment, material disposal and drinking water (Eggert, 1994). These permits usually have to be renewed periodically and constitute an active and permanent responsibility. In this context, water quality controls are essential to be in compliance with environmental policies, at the federal and state level. For instance, one of the most serious problems facing mining is the creation of acid rock drainage by the interaction of solid wastes and surface or groundwater. Metals dissolved in the acidic water can contaminate drinking water if they get to be discharged into surface or groundwater. Most common water contaminants in mining are:
- **Solid particles**: generated from drainage of the operation area, pockets of erosion and effluents.
- **Oils and fats**: usually present at mechanical workshops, mining equipment lubrication, supplying fuels, spills, equipment, and vehicle washing areas.

- **Acids**: can come from the same reservoir, waste disposal, transport and handling of acids during the process.
- **Organic contaminants**: sanitary facilities, canteens and laundry detergents.
- **Metals:** The presence of metals is associated with the production of acid drainage.
- **Cyanides**: used for the leaching of gold.
- **Alkalis**: when caustic soda is used to raise the pH in the flotation of minerals.
- **Salts**: derived from the geological substrate itself or in reagents.
- **Compound of nitrogen and phosphorus**: by the process of flotation of minerals.
- **Radionuclides**: Present in uranium, thorium and radium-226.

There are also other sources of potential water pollution, such as the generation of wastewater in mines:
- **Mine water:** Contamination of mine wastewater have: (a) concentrations of suspended solids outside the limits; b) PH below or above standards; c) high concentrations of dissolved metals including lead, copper, zinc, iron, manganese; d) high concentrations of total metals, such as lead, copper, zinc, iron, manganese, arsenic, mercury, selenium, nickel and cadmium.
- **Acid drainage from mines:** Resulting from the interaction of metals, air and water in an oxidizing environment. Once created, acid waters can involve other minerals to form solutions containing cadmium and arsenic. Acid drainage from mine can affect plant and aquatic life if discharge to the surface or groundwater. Acid mine drainage can be divided into: a) Very acidic: they have a net acidity range greater than 300 mg/l as CO3Ca; b) Moderately acidic: have a net acidity range between

100 and 300 mg/l as CO3Ca; c) **Weakly acidic:** with a net acidity range between 0 and 99 mg/l as CO3Ca; d) **Weakly alkaline:** with an alkalinity Net less than 80 mg/l as CO3Ca and e) **Strongly alkaline:** with net alkalinity greater than 80 mg/l as CO3Ca.

- **Tailings:** Floods resulting from tailings dam failures are the source of serious environmental damage. Tailings are metallurgical effluents from the treatment of minerals and are composed of water and solids, which includes: a) Suspended solids; b) Small concentrations of metals in solution; c) Chemical reagents used in the process: Sodium cyanide, dithiophosphates, xanthates, chromates, sulphites, copper sulphate, zinc sulphate, fatty acids, alcohols, oils, and pH modifiers such as lime, sodium hydroxide, limestone, sodium carbonate, sulfuric acid, inter alia.
- **Wastewater:** from the beneficiation plant, which contains metals, heavy metals and chemical reagents used in mineral treatment.
- **Acid waters from debris deposits:** With high solids content in suspension, high acidity or concentration of dissolved metals.
- **Domestic wastewater:** From the human activity, such as food preparation, garbage, sanitary use, cleaning products, personal cleanliness and laundry.

Mine operations must prepare and execute a water monitoring and maintenance plan to deal with potential contamination and ensure all risks of contamination to the environment and public health are eliminated by an adequate treatment of drainage water before being released. The main objective is to characterize and control of the following variables of the water, each of them must be analyzed in specific form:

- **PH**
- **Cyanides**
- **Oils and fats**

- **Suspended solids**
- **Heavy metals:** iron, mercury, lead, copper, zinc, manganese, selenium, nickel, cadmium and others
- **Arsenic**
- **Thermotolerant coliforms, biochemical oxygen demand, chemical oxygen demand, and temperature.**

Water Quality Characteristics

Water is a colorless, odorless and insipid substance. Water has chemical, physical, microbiological and biological characteristics (Fig. 2.1). These characteristics will determine whether water can be suitable for consumption or not. Each mine site will develop analytical methods to conduct sampling for field or laboratory analysis to determine water parameters at any significant discharge point.

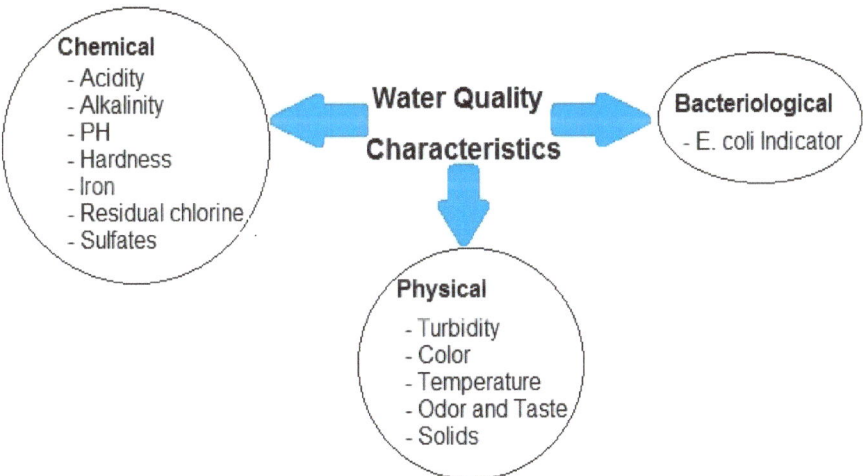

Fig. 2.1 Water Quality Characteristics

1. **Physical characteristics**

Consider by its appearance, which can be perceived by the senses. Physical characteristics of water are:

- **Turbidity:** Turbidity is an optical effect caused by light rays passing through water containing suspended and colloidal material. It is an important characteristic for aesthetic and health reasons.
- **Color:** Color in water is expressed in platinocobalt (units of color). Color is determined by comparing the sample with previously measured patterns. Just by its appearance, colored water should be discarded to be taken (even though it may have no risk). Colored water can indicate the presence of potentially dangerous organisms in the water.
- **Temperature:** Water temperature is important because it directly affects the physical properties of water, its chemical reactions, density, surface tension, salinity, the solubility of the dissolved gases, thermal capacity, viscosity, expands flavors and odors of water and determines the development of organisms in it. It is determined with a thermometer and expressed in degrees centigrade. The temperature of the water in the natural state in the world ranges from 0 to 35 °C.
- **Odor and Taste:** Taste and odor are part of our senses. The dissolved organic impurities produce odors and flavors that tasting and smelling detect. Odors in the water are due to very small concentrations of volatile compounds, coming from soil and mold, putrefaction or petroleum derivatives. In the case of surface water, the odor is caused by plankton. Human taste perception detects metallic salts such as copper, zinc or iron as they provoke bitter, salty, sweet and bitter tastes.
- **Solids:** Are the residues at the end of the evaporation of the water and the drying of the residue to a constant weight of 103 °

C to 105 ° C. These are classified as sedimentable, suspended and filterable solids.

2. Chemical characteristics

It is related to the microbial and chemical transformations that affect the chemical characteristics of the water. They are usually the product of the material in contact that has been either soils or rocks. These characteristics are:

- **Acidity:** The acidity of water is a measure of the total amount of acidic substances (hydroxyl ions) present. This acidity occurs when pH = 4.3 (strong acidity that produces bad odors and flavor) and pH = 8.3 (weak acidity that produces corrosion in steel, concrete, others). The acidity disappears in the water when pH is greater than 8.3.
- **Alkalinity:** Alkalinity is defined as the ability to neutralize water when acid is added. It is expressed as milligrams per liter of ($CaCO_3$) calcium carbonate. It is the ability of water to react with an acid. Total alkalinity is often due to the presence of Hydroxide Alkalinity [OH^-], Bicarbonate Alkalinity [HCO_3^-] and Carbonate Alkalinity [CO_3^{2-}]. If pH of water is higher than 8.3 there is a presence of hydroxides in water and some carbonates. If pH of the water is between 4.3 and 8.3 there is a presence of bicarbonates and carbonates. If pH of water is less than 4.3, there is no alkalinity.
- **PH:** Water always ionizes in small proportions, both in hydrogen ions and in hydroxyl ions. PH is the way to measure the hydrogen ion or the hydroxyl ion. Its values are between 6.5 and 9.0. Neutral pH = 7. The acid pH when it is less than 7. The alkaline pH when it is greater than 7. It is found by colorimetric or potentiometric comparison and is expressed in pH units.

- **Hardness:** Water hardness is due to the presence of Total dissolved solids. It represents the total concentration of calcium and magnesium ions mainly. It is temporary hardness by the presence of calcium carbonates and bicarbonates of magnesium and hydroxides. It is permanent hardness when there is calcium, manganese, manganese and iron. Hardness is expressed in mg / L CaCO3, and classifies water in "soft" or "very hard"
- **Iron:** Presence of iron in water has a corrosive effect that alters water properties: odor, taste and aesthetics. Iron Permissible Value is less than 0.3 mg / lt.
- **Residual chlorine:** The chlorination of water allows for its consumption since through disinfection all bacterial agents are eliminated. Chlorine can be applied as gas or in solution and mixed with hypochlorites.
- **Sulfates:** Sulfates come from soils that are rich in minerals. Its admissible value is less than 250 mg / lt.

3. **Bacteriological characteristics**

Indicator organisms such as Escherichia coliforms (E. coli) or Pseudomonas aeruginosa can be present in water, that if ingested for humans may suggest the presence of sewage.

II. WATER RIGHTS

State and Federal governments are involved in water rights in the U.S. Distinct laws are developed based at a great extent on economic characteristics in relation with water, geographical conditions, and land ownership.

Water rights in most of the arid western states (Alaska, Arizona, Colorado, Idaho, Montana, Nevada, New Mexico, Utah, and

Wyoming) fall under the **"prior appropriation doctrine"**, which refers to the water rights given to the person who first puts water to beneficial use. States under this law had validated the principle as well as court decisions. The **"Riparian law"** is the standard adopted by most of the eastern and Midwestern states. Riparian is related to a land located on the banks of a stream or river. The riparian rights give the water right to the landowners of the land to reasonable use (and trough permit application) the water on their lands. The remaining states operate under a combination of prior appropriation and riparian law basis.

Doctrine	States
Pure Prior Appropriation	Alaska, Arizona, Colorado, Idaho, Montana, Nevada, New Mexico, Utah, and Wyoming
Pure Riparian	Missouri, Tennessee, Ohio, West Virginia, Louisiana, Maine, New Hampshire and Vermont
Prior appropriation, formerly riparian	Washington, Oregon, North Dakota, South Dakota, Kansas and Texas
Regulated Riparian	Minnesota, Iowa, Wisconsin, Illinois, Michigan, Kentucky, Pennsylvania, New York, Arkansas, Mississippi, Alabama, Georgia, Florida, South Carolina, North Carolina, Virginia, Maryland, Delaware, New Jersey, Connecticut, Rhode Island, Massachusetts
Mixed Riparian-Prior Appropriation	California, Nebraska and Oklahoma
Other	Hawaii

III. AGENCIES

In any regulatory environmental framework, there are fundamental responsible parties for its compliance. Laws and regulations on water are dictated by federal and state agencies and observed by a wide range of environmental agencies. Below is a list of relevant agencies involved in regulating water in the mining sector in the United States, which are classified in federal agencies and environmental agencies.

1. Federal agencies

Mining operations are regulated by the environmental laws managed by this Federal agency whose responsibility is the management of the nation's resources.

a) **The United States Environmental Protection Agency (EPA)**

EPA is an agency of the federal government of the United States, whose primary mission is to protect human health and the environment with the regulation approved by the Congress. It was established in 1970 by President Richard Nixon, who in turn created 10 EPA Regions (Fig. 2.2), each of which is responsible for the management of EPA programs:

- Region I: States of Connecticut, Maine, Massachusetts, New Hampshire, Rhode Island, and Vermont.
- Region II: States of New Jersey and New York. Puerto Rico and the U.S. Virgin Islands.
- Region III: States of Delaware, Maryland, Pennsylvania, Virginia, West Virginia, and the District of Columbia.
- Region IV: States of Alabama, Florida, Georgia, Kentucky, Mississippi, North Carolina, South Carolina, and Tennessee.

- Region V: States of Illinois, Indiana, Michigan, Minnesota, Ohio, and Wisconsin.
- Region VI: States of Arkansas, Louisiana, New Mexico, Oklahoma, and Texas.
- Region VII: States of Iowa, Kansas, Missouri, and Nebraska.
- Region VIII: States of Colorado, Montana, North Dakota, South Dakota, Utah, and Wyoming.
- Region IX: States of Arizona, California, Hawaii, Nevada, the territories of Guam and American Samoa, and the Navajo Nation.
- Region X: States of Alaska, Idaho, Oregon, and Washington.

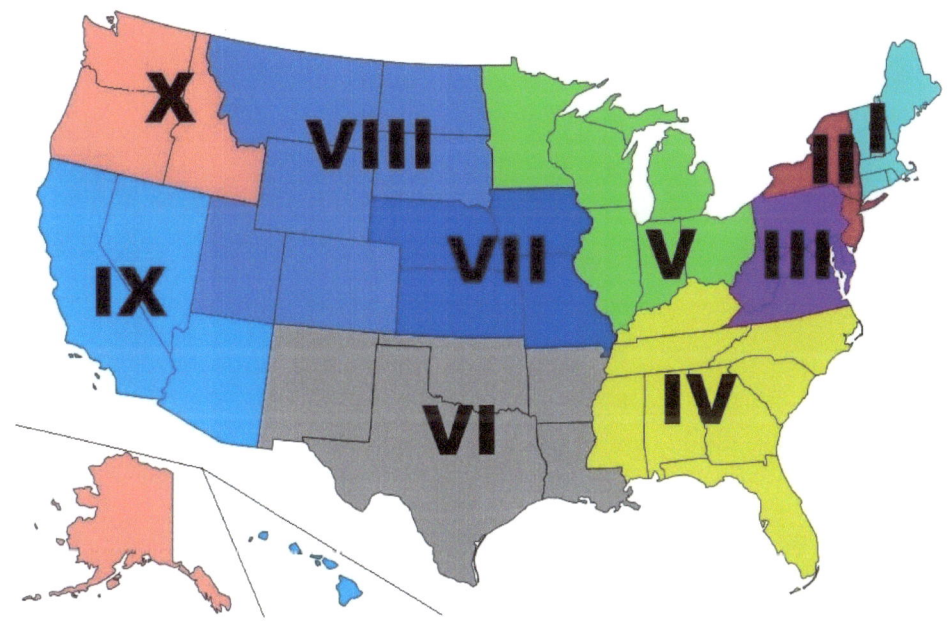

Fig. 2.2 EPA Regions

Its purpose is to ensure:
- Protection to all Americans from human health and their environmental significant risks
- Reduction environmental risk is based on scientific information

- Enforces federal laws that protect natural resources, including air, water, and land
- All parts of society have accurate information to participate.

Their programs include:
- EPA Safer Choice
- Safer Detergents Stewardship Initiative
- Water Sense
- Drinking water
- National Pollutant Discharge Elimination System
- Environmental Education

Related legislation
- 1948: Water Pollution Control Act PL 80-845
- 1965: Water Quality Act PL 89-234
- 1966: Clean Waters Restoration Act PL 89-753
- 1970: Water Quality Improvement Act PL 91-224
- 1972: Federal Water Pollution Control Amendments of 1972 PL 92-500
- 1974: Safe Drinking Water Act PL 93-523
- 1977: Clean Water Act PL 95-217
- 1987: Water Quality Act PL 100-4
- 1996: Safe Drinking Water Act Amendments of 1996

b) The Bureau of Land Management (BLM)

BLM is an agency within the Department of the Interior, manages diverse landscapes and resources across the nation and administers more than 1,001,000 km2 of public lands in the United States. The BLM was established in 1946 when the young nation began acquiring more lands. Then President Harry S.

Truman merged the General Land Office and the Grazing Service agency creating the BLM.

BLM Mission

"To sustain the health, diversity, and productivity of the public lands for the use and enjoyment of present and future generations".

The BLM public lands regions include: The National Office in Washington DC, Arizona, Alaska, Colorado, California, Eastern States, Montana, Idaho, Nevada, New Mexico, Oregon, Utah, and Wyoming.

Soil, Water, Air Program

This program includes both surface water and groundwater sources. The objective is to assess and restore water quality conditions and to manage water resources on public lands. Clean and quality supplies of water are necessary to incentive healthy watersheds, provide safe habitat for fish and wildlife, maintain drinking water sources, allow for the safe recreational use of our surface water and maintain healthy plant communities.

The goals of the BLM's 5-year strategy program (Ford & McCormick, 2015) are:
- *Reducing the discharge of traditional pollutants and sediments into water resources.*
- *Incorporating collaborative, regional watershed assessments into BLM planning efforts to understand potential impacts to watersheds from land decisions.*
- *Improving water quality monitoring. Maintaining their technical expertise.*

- *Implementing protection, maintenance, restoration of the health and diversity of ocean, rivers and Great Lakes ecosystems.*

c) The United States Forest Service (FS)

FS is an agency of the U.S. Department of Agriculture that administers the 154 national forests and 20 national grasslands, around 780,000 km2.

How does forest supply water?
Forest captures and holds snow in winter. Forest soils absorb rain and snow acting as big sponges, a natural filter, replenishing underground aquifers. Forests in the United States provide drinking water to more than 180 million people (Furniss, 2010). The National forest supplies a reliable clean and cold water to about 20 percent of the American population. That is to say, 66 million people depend on the national forest as their water source. Forest Service administers the largest source of water in the U.S. with field specialists dedicated to promote and improve sustainable, healthy watersheds essentials to people and ecosystems (U.S. Department of Agriculture, 2017)

Forest Service mission:
- *Caring for the land and serving people.*
- *Achieve quality land management under the sustainable multiple-use management concept to meet the diverse needs of people.*

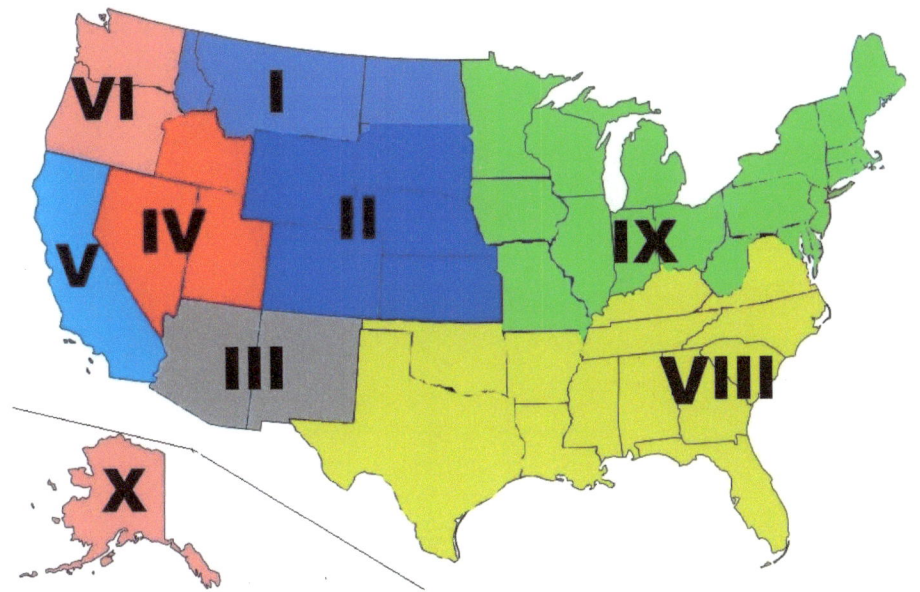

Fig. 2.3 United States Forest Service Regions

The Fig. 2.3 above show the nine regions in the USDA Forest Service:
- Region I: based in Missoula, Montana.
- Region II: based in Golden, Colorado.
- Region III: Based in Albuquerque, New Mexico.
- Region IV: based in Ogden, Utah.
- Region V: based in Vallejo, California.
- Region VI: based in Portland, Oregon.
- Region VIII: based in Atlanta, Georgia.
- Region IX: based in Milwaukee, Wisconsin.
- Region X: based in Juneau, Alaska.

d) The United States Fish and Wildlife Service (FWS)
FWS is an agency of the federal government of the U.S. Department of the Interior. FWS is dedicated to protect fish and

wildlife and their habitats. The Bay Delta office (From FWS) is an agency of the Department of Interior, which conserves and protects fish, wildlife, and plants. Also, the FWS's Bay Delta Office coordinates with other federal agencies on the potential impact of federal projects (U.S. Fish & Wildlife Service, 2013)

Fish and Wildlife Service Mission
- *"Working with others to conserve, protect, and enhance fish, wildlife, plants and their habitats for the continuing benefit of the American people."*

e) **The Bureau of Reclamation (BR)**
BR is a federal agency under the U.S. Department of the Interior, manages federal water projects in sixteen western states. Projects as water resource management, delivery, storage irrigation and water supply. More than 31 million Americans are supplied with water from the BR source (Bureau of Reclamation, 2016)

BR mission
- *Meet new water needs and balance the multitude water demands of the West while protecting the environment and the public's investment in these structures.*
- *Fulfilling the water delivery obligations, water conservation, water recycling and reuse, and developing partnerships with our customers, states, and Native American Tribes.*

List of reclamation projects:
- Animas-La Plata Water Project
- Boise Project
- Central Arizona Project Aqueduct
- Central Utah Project

- Central Valley Project
- Colorado-Big Thompson Project
- Colorado River Storage Project
- Columbia Basin Project
- Elwha River Dam Removal Project
- Fryingpan-Arkansas Project
- High Plains Cooperative Pilot Project
- Huntley Project
- Klamath Project
- Moon Lake Project
- Navajo-Gallup Water Supply Project
- North Platte Project
- Pojoaque Basin Regional Water System Project
- Project Skywater
- Rio Grande Project
- Shoshone Project
- Sierra Cooperative Pilot Project
- Strawberry Valley Project
- Washita Basin Project
- Yuma Project

The BR regions include Lower Colorado Region, Mid-Pacific Region, Pacific Northwest Region and Upper Colorado Region.

f) The Federal Emergency Management Agency (FEMA)

FEMA is an agency of the United States Department of Homeland Security, administers the National Flood Insurance Program (NFIP), disaster planning and recovery programs. FEMA provides funding and technical assistance to states and communities and supplement floodplains data and flood critical maps to better flooding responses. FEMA coordinates with states and communities regarding disaster response to such event occurring nationwide (U.S. Department of Homeland Security, 2017)

Regional are presented in Fig. 2.4:
- Region I: Boston, MA Serving
- Region II: New York, NY Serving
- Region III: Philadelphia, PA Serving
- Region IV: Atlanta, GA Serving
- Region V: Chicago, IL Serving
- Region VI: Denton, TX Serving
- Region VII: Kansas City, MO Serving
- Region VIII: Denver, CO Serving
- Region IX: Oakland, CA Serving
- Region IX: PAO Serving:
- Region X: Bothell, WA Serving

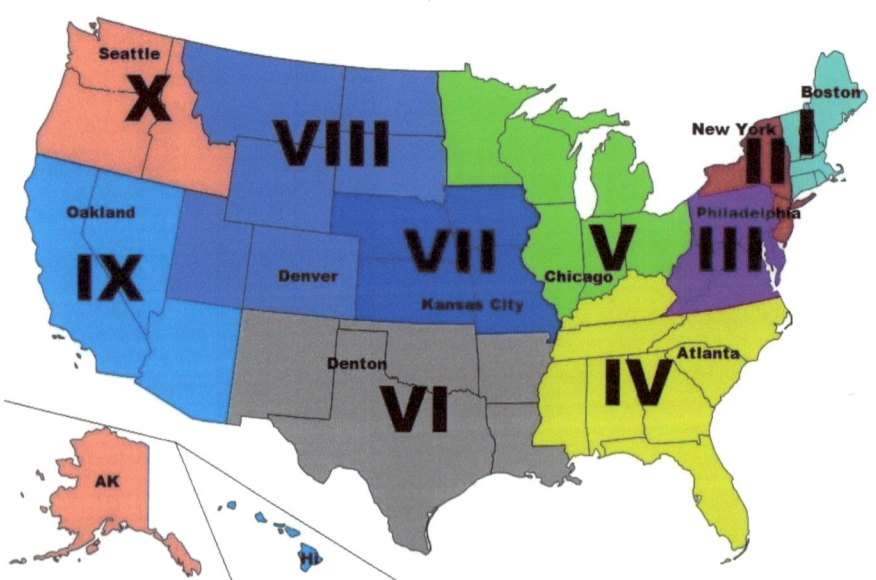

Fig. 2.4 Federal Emergency Management Agency Regions

g) U.S. Army Corps of Engineers (USACE)
USACE is a U.S. federal agency of the Department of Defense. The U.S. Army Corps of Engineers is commanded by civilian and military personnel (USACE, 2017)

Environmental missions:
Restoring degraded ecosystems, constructing sustainable facilities, regulating waterways, managing natural resources, and cleaning up contaminated sites from past military activities.

h) United States Geological Survey (USGS)
USGS is a scientific agency of the United States government that provides "impartial information about the health of our ecosystems and environment, the natural hazards that threaten us, the natural resources we believe, the impacts of climate and land-use change." Scientists of the USGS analyze the landscape of the country in order to understand their essential features and hazards threaten our natural resources, revealing their research to the public. The USGS conducts their investigations in a scientific mode, by collecting, analyzing, and disseminating impartial information and hydrologic data.

USGS mission
- *Collect and disseminate reliable, impartial, and timely information that is needed to understand the Nation's water resources.*

i) The River Network
The River Network's mission is *"to empower and unite people and communities to protect and restore rivers and other waters that sustain the health of our country"*. The River Network is a national

watershed protection movement founded in 1988. The staff is headquartered in Portland, Oregon, with field offices in Maryland, North Carolina, Ohio, and Utah. (River Network, 2017)

River Network Mission
- *Empowers and unites people and communities to protect and restore rivers and other waters that sustain all life.*
- *A future with clean and ample water for people and nature, where local caretakers are well-equipped, effective and courageous champions for our rivers.*
- *Everyone should have access to affordable, clean water and healthy rivers.*

2. Environmental agencies

Environmental agencies are organizations recognized as non-profit corporations under U.S. Internal Revenue Service Code, Section 501 C-3. Currently, exist more than 1500 of them. To mention only the most influential ones in regards to the organization size, under three categories:

a) **Larger mainstream associations**: with similar viewpoints
- **Ducks Unlimited:** working on the conservation of wetlands, associated upland habitats, wildlife and people (Ducks Unlimited, 2017)
- **Earth Watch:** its mission is to engage people worldwide in scientific field research and education to promote the understanding and action necessary for a sustainable environment (Earthwatch Institute, 2017)

- **Environmental Defense Fund:** working on global warming, ecosystem restoration, oceans, and human health, and advocates through the science, economics, and law to find environmental practical solutions (Environmental Defense Fund, 2017)
- **National Audubon Society:** an organization working to conserve and restore of natural bird's ecosystems and others, their habitats for the benefit of humanity and the earth's biological diversity (National Audubon Society, 2017)
- **National Parks and Conservation Association**: its mission is to protect and enhance America's National Park System for present and future generations (National Parks and Conservation Association, 2017)
- **National Wildlife Federation:** that works inspiring Americans to protect wildlife for the children's future (National Wildlife Federation, 2017)
- **Natural Resources Defense Council:** which purpose is to safeguard the Earth and all that depend on them (people, plants, animals, natural systems). (Natural Resources Defense Council, 2017)
- **The Nature Conservancy:** its mission is the conservations of lands and waters because all life depends on them (The Nature Conservancy, 2017)
- **The Sierra Club:** one of the oldest, most effective and powerful conservation organizations to make changes in government and corporations in the USA (Sierra Club, 2017)
- **The Wilderness Society:** its mission is to protect wilderness and inspire Americans to care about it (The Wilderness Society, 2017)
- **World Wildlife Fund (Canada & USA):** An organization who dedicates to promote science, working thought wilderness

preservation and not increasing of humanity's footprint on the environment (WWF, 2017)
- **World Wide Fund for Nature (International):** An organization that promotes to stop the degradation of the planet's health and to build a future where human beings live in harmony with the ecosystem and using of renewable natural resources.

b) <u>Mid-sized mainstream and regional clubs</u>: use environmental issues as the key to force a fundamental transformation of the present mode of society by drastically changing the contemporary lifestyle.
- **Earth First:** a radical environmental group, which works for the protection of wilderness and wildlife (Earth First, 2017)
- **Friends of the Earth:** its mission is to work on environmental and social justice, human dignity, respect for human rights to secure sustainable societies (Friends of the Earth, 2017)
- **Greenpeace:** working using a zero violence, creative confrontation to show global environmental problems. Their goal is to ensure the ability of Earth to nurture life in all its diversity (Greenpeace USA, 2017)
- **Sea Shepherd Society:** its mission is to stop the destruction of habitat and slaughter of wildlife in the world's oceans in order to conserve and protect ecosystems and species (Sea Shepherd Society, 2017)

c) <u>Mainly small or single-issue groups:</u> whose thrust is either mainly or fully directed towards the mining industry:
- **Alliance for the Wild Rockies:** focused on the integrity of the Wild Rockies Bioregion through citizen empowerment, using the conservation biology, sustainable economic models and environmental law (Alliance for the Wild Rockies, 2017`)

- **Center for Alternative Mining Development Policy:** provides technical and educational information on the impacts of mining to rural and Indian communities throughout Wisconsin.
- **Clark Fork Coalition:** working on the Clark Fork basin, that people, fish, and wildlife have clean water and healthy rivers (Clark Fork Coalition, 2017).
- **Colorado Mining Action Project:** its goal is to protect water from uranium mining and protect the rights to public involvement.
- **Project Environment Foundation**: committed to making every donation, gift or bequest count because the best long-term environmental outcomes will only be achieved by funding activities which are integrated with the bigger natural resource management picture for the area.
- **Save Lake Superior Association:** organized to stop the dumping of taconite tailings into Lake Superior by Reserve Mining, whose tailings were contaminating the water and aquatic life and affecting human life (Save Lake Superior Association, 2017)
- **Southwest Research and Information Center (SRIC):** working on project to take care health of people and communities, protect natural resources, and ensure citizen participation, secure environmental and social justice now and for future generations (SRIC, 2017)

IV. SURFACE WATER

1. Introduction

Surface water is the water in rivers, streams, creeks, lakes, oceans, and reservoirs. Surfaces waters are important because are used to supply potable water to approximately 200 million people in the United States (EPA 2004), other uses like irrigation and by the thermoelectric-power industry to cool electricity-generating equipment and supply a habitat for fish and amphibians, birds and reptiles. For the mining industry, surface water represents 27% of all new water used, being groundwater the rest.

For example, Figure 2.5 shows a satellite image of a copper mine between two lakes, located in Salt Lake County, Utah.

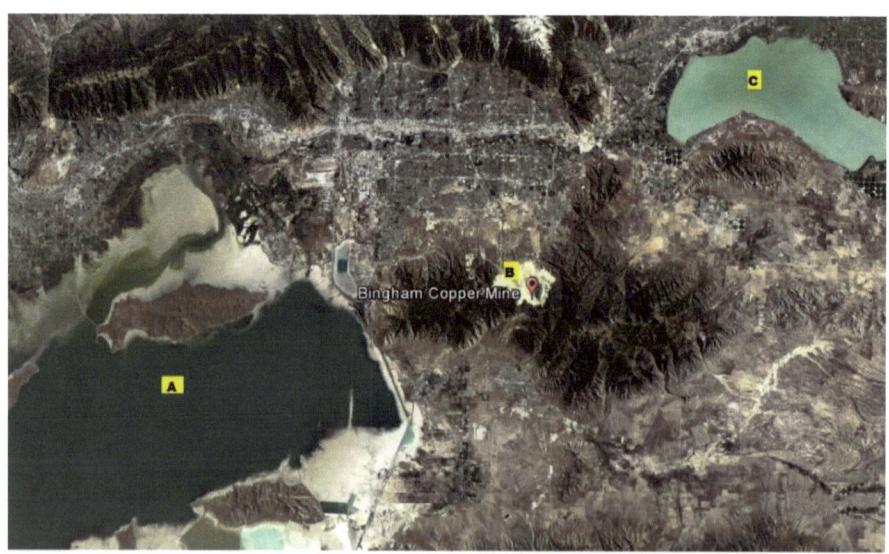

Fig. 2.5 Satellite image of Bingham Canyon Mine located adjacent to the Great Salt Lake and Utah Lake (Google Earth Pro, 2017). Key points of interest are shown by the letter annotations: A: The Great Salt Lake, B: Open Pit Mine, C: Utah Lake.

Mining activities produce changes in the surface such as leveling off the ground, diversion of runoff and the laying of wastes that increase or decrease infiltration into groundwater. The biggest problem, usually devastating, that occurs in metal mines is the Acid Mine Drainage (AMD), which causes substantial short and long duration damage to all surface waters downstream such as lakes, lagoons, rivers, streams, animals in these waters.

The mining impacts that affect water quality are:

- Acid Mine Drainage (AMD)
- Heavy metals and leaching
- Chemical Contamination
- Erosion and sedimentation

2. Regulations

In the USA, the laws for protecting surface water quality are Safe Drinking Water Act (SDWA) and Clean Water Act (CWA), both are administrated by the U.S. Environmental Protection Agency (EPA).

2.1. Safe Drinking Water Act (SDWA)

The Safe Drinking Water Act (SDWA, codified at 42 USCA Secs. 300f et seq.) is the federal law since 1974 that regulates public drinking water systems throughout the United States.

During its history, the law has been twice amended in 1986 and 1996 working in many actions to protect drinking water and its sources like rivers, lakes, reservoirs, springs, and groundwater wells.

Public water systems, according to the law, serve at least 25 people per day for 60 days of the year, so it doesn't regulate for private wells that serve less than 25 individuals.

The applicability of this law distinguishes water systems based on their type and size:

- Community Water System
- Non-Community Water System
- Non-Transient Non-Community Water System
- Transient non-community water system

EPA and SDWA set standards to protect against contaminants that may be found in drinking water through the implementation of programs, they work together making sure these standards are met.

The most common threats to safe drinking water are the use of pesticides, incorrect disposition of chemical, human threats, animal waste injected underground and natural contamination.

At first, SDWA focused primarily on treatment as the means of providing safe drinking water at the tap. After the 1996 amendments, SDWA ensures providing quality safe drinking water from source to tap. The amendment highlights were:

- Annual Consumer Confidence Reports
- Cost-Benefit Analysis by US EPA for every new standard
- Funding for improvements or to help systems assess and protect the source water
- Microbial Contaminants and Disinfection Byproducts to strengthen protection for microbial contaminants
- Water system operators must be certified
- Public Information & Consultation rights

- Small water systems are given special consideration and resources under SDWA
- Every state must assess its sources of drinking water

There are more than 170,000 public water systems that provide water to people in the USA. The Roles and Responsibilities are:

- **US EPA, states, water systems, tribes and the public**: responsibility for making sure to provide safe drinking water
- **SDWA**: implementation of barriers against pollution, including source water protection, treatment, integrity for the distribution system and finally public information.
- **Public water systems**: controlling amount of tap water contaminants so they do not exceed regulations.
- **Water systems**: treat the water, and must test their water frequently for specified contaminants and report the results to the states. It requires notification to customers if standards are not met.
- **Public**: help to set priority goals to the suppliers of water, working on funding decision makers and system improvements, and establishing programs to protect drinking water sources.

Safety drinking water standards Process:

- **First:** Identification of contaminants in drinking water for further study. Determination of contaminants to potentially regulate by US EPA.
- **Second**: Determination of a maximum contaminant level goal in drinking water for contaminants (no expected risk to health) regulate by US EPA.
- **Third**: Specification of the maximum permissible level of a contaminant in drinking water which is delivered to any user of

a public water system that may be achieved with the use of the best technology, treatment techniques or ways to remove contaminants.

2.2. Clean Water Act (CWA)

The Clean Water Act (CWA, 33 USCA Secs. 1251 et seq.) known as "CWA" or "Act" is the most important regulation law about the environmental aspect of water. The CWA is the cornerstone of surface water quality protection in the United States establishing a national standard and programs that regulate the discharges of pollutants into the waters of the USA and regulating quality standards for surface waters. The purpose of the CWA is to restore most nation's waterways of USA into swimmable, fishable and drinkable.

Everything goes back to the year 1948 where the Clean Water Act was enacted and called the Federal Water Pollution Control Act, then expanded in 1972 and unanimously changed its name to "Clean Water Act". The Federal Water Pollution Control Act (FWPCA) of that year marked a major event of state policy for the federal government.

According to the law, the Environmental Protection Agency (EPA) is the regulatory authority to effectively meet this goal development. It uses many regulatory and non-regulatory tools to fully reduce direct pollutant discharges into waterways, increase funding for municipal wastewater treatment facilities, and manage polluted runoff to restore and maintain chemical, physical and the waters of the nation.

For this it was necessary:
- Establish national effluent limits based on technology for polluting sources.
- Write permissions for downloads.
- Coordinate, approve, guide and apply water quality programs throughout the country.

In the USA, the Clean Water Act establishes water quality standards (WQS) that includes all water like wetlands, estuaries, rivers, and lakes (EPA, 2003). WQS has three components:
- **Designated uses for the body of water:** that the state and the federal government have established the surface water body conditions.
- **Criteria with quantitative water quality measures:** levels of a pollutant, water quality characteristics and descriptions of waterbody conditions.
- **Ant degradation policies**: Improvement policies to maintain the criteria mentioned above, such as the use of anti-degradation measures.

CWA has regulatory standards for effluents, categorizes the specific use for water bodies and operates the NPDES permit system for wastewater treatment facilities.

On the other hand, each state must establish technological standards and discharge limits so that the permitted entities reach broad federal water quality standards. If this does not happen, the EPA is authorized to assume control of the state.
In the USA, the water discharges are classified in:

- **Point sources of pollution**: Are the discharges related to a wastewater treatment plant or a stormwater drainage system, discharging a river into the stream.
- **Non-point sources (NPS) of pollution**: Are diffuse, include runoff from the earth's surface and the same atmospheric deposition.

These days, EPA is engaged in efforts to reduce pollution levels through the treatment of non-point sources, as these are not regulated under the CWA, and are largely responsible for a large part of water pollution of the nation. EPA works actively to provide states with technical and financial assistance to implement programs to reduce NPS contamination levels.

The following are two of the fundamental programs administrated under the Clean Water Act (CWA): The National Pollutant Downloads Elimination System (NDPES) and The Clean Water Act- Section 404 Permit Program.

a. The National Pollutant Discharge Elimination System (NPDES) Program

NPDES is an American permit program that seeks to reduce water pollution at point sources that discharge pollutants into the water.

This program was established in 1972 by the Clean Water Act and is the largest enforcement mechanism of the Act. It works with EPA-regulated state governments to carry out many aspects of permitting, administrative and program implementation.

It is totally prohibited to discharge any point source of contamination to navigable waters, without the permission of

NPDES (the regulations are in 40 C.F.R. Pam 121-125). Point sources such as pipes, gutters, channels, tunnels, trenches as the case of industrial, municipal and other facilities such as discharges of mining activities that discharge directly to surface waters.

In contrast, this does not apply to individual dwellings that are connected to a municipal system, use a septic system or do not have a surface discharge. These activities do not need a NPDES permit.

What is this program about?
A NPDES permit must be obtained for any discharge of contaminants from a point source into United States waters. Figure 2.6 is an example of NPDES permit for the city of Martinsburg. Limits of pollutant discharges are spelled out in the permit and require self-monitoring of compliance. In general, all activities that discharge from contaminants to surface waters are potentially subject to the NPDES program.

Fig.2.6. Tuscarora Creek NPDES Permit, Martinsburg, Virginia, USA

It should be emphasized that there is currently no single list of pollutants regulated by the Act, only general prohibitions to limit the discharge of toxic or dangerous pollutants. Any discharge of natural groundwater from a mine containing trace metals and solids will require a NPDES permit. The NPDES permit may be issued by the state or EPA prior to any discharge.

Authorization for states, tribes, and territories is obtained through a process that is defined by Clean Water Act (CWA). Submission of the following items is needed in order to receive this authorization:
- Letter from the Governor requesting review and approval,
- Memorandum of Agreement (MOA),
- Program description,
- Statement of legal authority (also known as an "Attorney General's Statement" or "AG Statement"), and
- The underlying state laws and regulations.

NPDES Permit Limits:
- **TBELs and WQBELs**: Permits should consider limits based on the Technology-based Effluent Limitations (TBELs) and limits that protect the standards of receiving water quality called Water Quality-based Effluent Limitations (WQBELs). The TBELs permits require a minimum level of treatment of contaminants for point source discharges based on available treatment technologies, permitting the discharger to use any control technique available to accomplish the goal. Through a process outlined by each state, it will be determined if the implementation of certain technology-based controls does not

suffice to meet water quality requirements. They rank the list of contaminated water priorities and establish total daily maximum loads (TMDL). The TMDL identifies the amount or property of a contaminant, from sources including a safety margin, that can be discharged to a body of water and still ensure that the body of water meets water quality standards.

- **Nutrient Permitting:** The contamination of nutrients in bodies of water is when the nutrient load exceeds the limit of presence in water bodies. The effects are accelerated eutrophication (due to the presence of nitrogen and phosphorus transported by rainfall) and low water quality. The point sources that discharge nutrients are municipal and industrial wastewater treatment plants, concentrated animal feeding operations (CAFOs), municipal sewage systems (MS4s), rainwater associated with industrial activity, among others.
- **Recreational Water Quality Criteria Limits:** Applicable to continuous dischargers (POTWs) based on water quality standards for pathogens and pathogens indicators associated with fecal contamination in primary contact recreational waters. There are recommendations for two bacterial indicators of fecal contamination, enterococci, and E. coli.
- **TMDL Permitting:** A TMDL is the maximum amount of a contaminant that a body of water can receive and still meet water quality limits. They consist of a waste load set (WLAs) for point sources, load assignments for non-point sources, a safety margin and possibly a reserve allocation.
- **Watershed-based Permitting:** Permit that uses watersheds, through which it focuses on the approach of all the stressors within a hydrologically defined watershed, and not only on the individual polluting sources in a discharge. The permits are varied, ranging from the synchronization of permits within a

watershed to the development of effluent limits based on water quality.

- **Whole Effluent Toxicity (WET):** Total Effluent Toxicity (WET) means the aggregate toxic effect of an aqueous sample (such as the discharge of wastewater from the entire effluent) as measured by the response of an organism upon exposure to the sample (like lethality, deteriorated growth, or reproduction). EPA WET tests replicate the total effect of environmental exposure from aquatic life to toxic pollutants in an effluent. WET testing is a key component in implementing water quality standards under the NPDES permit program.

b. Clean Water Act- Section 404 Permit Program

Mining activities and projects include dredging or landfill discharge into US waters and wetlands (Figure 2.7). A wetland is an area flooded by surface or groundwater with sufficient frequency and duration where vegetation is maintained. Prior to any construction or project, a CWA Section 404 permit is required to be implemented.

Fig. 2.7. Landfill discharge into US waters

This section indicates that the government is authorized to deny or restrict the program if a discharge of dredging or filler materials in such area will have an unacceptable harmful effect on the habitat or aquatic wildlife.

The government is authorized to issue permits if:
- Activities cause only minimal adverse effects when carried out separately, have only a minimal cumulative effect on the environment and compensation will be provided for all remaining unavoidable impacts.
- Activity demonstrates that measures have been taken to avoid impacts on wetlands, streams and other aquatic resources.

The permit review is administered by:
- **EPA**: responsible for compliance with discharges that are not allowed
- **Army Corps of Engineers of USA**: current issue of permits, individuals and general. Ensure compliance with permit conditions.

Implementation of the Section 404 Program:
- **Prevention:** The discharge of dredging or filler materials into the United States waters should be avoided. Determine if the activity depends on being in a body of water or being adjacent to it.
- **Minimization**: Changes on the project designs are mandatory in order to minimize impacts.
- **Compensation**: in which for each acre of wetland that is lost, at least one acre with equivalent function must be restored.

V. GROUNDWATER

1. Introduction

Groundwater is the water confine beneath the surface of the Earth, namely hills, mountains, arid deserts, plains and almost any place on the surface of the ground. Groundwater is part of the water cycle formed by rainwater and accounts for about 30% of the world's fresh water; melting ice accounts for almost 69% (ice sheets, mountain snow, glaciers) and water that filter through the bottom and can reach the subsoil (lakes, swamps, rivers) accounts for the other 1%. In many cases it is inaccessible and difficult to precise its location, perform measurements, and make a detailed description.

The age of groundwater depends on its depth, for example, if it can be only hours if found to be shallow, but deeper depths can be traced to be 100 years or reach thousands of years in case of being at great depths.

How does it arrive?
Precipitation (heavy rain) that falls on the ground will have several options to dilute, rainfall can go through the land and reach the seas, lagoons, rivers and can also hydrate plants, vegetation. Part of this water evaporates and returns to the atmosphere, and another fraction penetrates the soil, crosses the unsaturated zone until reaching the layer of water that is saturated soil or saturated rock material. In this area of saturated soil, the saturated groundwater moves slowly but it can be throw

streams, lakes, and oceans. The water table can be deep or shallow.

Groundwater is recharged when there is a leak in the supply systems or when more water is made than required by the crops.

The uses of this water are basically:
- Since groundwater is less likely to be contaminated, it generally conserves very good quality by being stored at great depths. It is protected from pollution, preserving its quality. It is used as drinking water (more than 50% use groundwater as drinking water nationwide)
- To be used in agriculture for the irrigation of crops.
- In the food industry.

Importance of groundwater:
- Regarding the environment: maintains the water level and flows to the rivers, lakes. During dry seasons (lack of rain), groundwater flows through the bottom of these bodies of water obtaining an ecosystem for animals and plants.
- Sustainable navigation: at dry seasons, groundwater is discharged into the rivers, allowing them to maintain high water levels.
- It does not require large investments in infrastructure and treatment.

2. Aquifers

Etymologically, aquifer comes from the Latin "aqua = water" and "fero = to carry" (See Figure 2.8). It is the name given to ground or underground rock through which groundwater can move, therefore it serves to store or produces naturally significant amounts of groundwater. They are generally composed of unconsolidated gravel, sand, permeable sedimentary rocks such as sandstones or limestones, volcanic rocks, and fractured crystals. The following geological formations are also considered aquifers: sandstones, limes, and dolomites, basalts, as well as fractured metamorphic and plutonic rocks are some examples of geological units considered aquifers. Water moves through these aquifer geological formations due to the permeability with which these large spaces are connected, and its subterranean water velocity depends on the size of these holes.

The rapidity of groundwater flow in the aquifer will depend on the characteristics of the aquifer. Usually, the direction of this movement is from high to low according to gravity. The movement of this sub-trench water will not end until it reaches another aquifer or other bodies of water, such as lakes, rivers, seas and the like.

Aquifers can be classified according to the hydrostatic pressure of water in:
- **Unconfined aquifers**: Aquifers that are in contact with the air at atmospheric pressure, and waterfalls from the surface directly to the aquifer. The level of the unconfined aquifer is the water table, and during mining activities, they can be reconfigured due to intensive well drilling.
- **Confined aquifers**: They are at high atmospheric pressures, occupying all the space that contains it, including the holes or pores of the geological formation in which it is found.

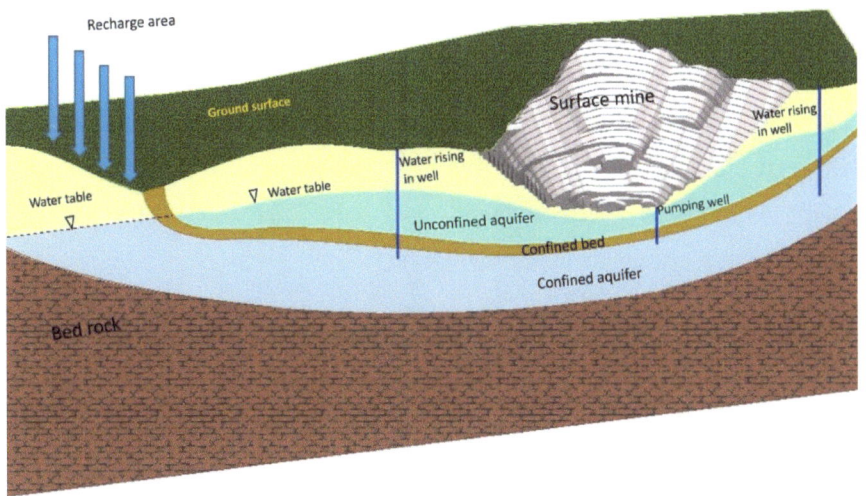

Fig. 2.8. Aquifer during mining operations

Physical characteristics of the aquifer:
- **Transmissivity**: The space must have holes in the form of pores or fractures. These spaces must be interconnected, so that permeability is achieved. When the spaces are not connected, it is said that the geological formation is impermeable. The transmissivity is measured by the amount of water that can be transmitted through the saturated aquifer waiting.
- **Storage Coefficient**: is the volume of water, per unit area and change in water height, which a permeable unit will absorb or release from storage. There is a direct relationship between the porosity and permeability of the aquifer with the amount of water to be stored or produced.

Groundwater located in the aquifers can be extracted through a drilled well and the help of a pump. Other aquifers naturally overflow to lakes and rivers due to natural pressures.

At present times, the estimated amount of water stored below the surface of the earth in form of aquifers is about 23 million cubic kilometers of total groundwater, and only about 22 million cubic kilometers are more than 50 years old.

Contamination of the aquifer

It is produced via filtration of pollutants from the surface soils into the ground until reaching the water table. Contamination of groundwater occurs through the washing of sources of poorly waterproofed contaminants such as urban or industrial waste dumps, septic tanks, underground hydrocarbon deposits, and materials produced by mining works (mineral enrichment, sterile waste streams), used salts for the thawing of roads. They can also be contaminated by the dripping and indiscriminate use of agricultural products such as chemical fertilizers (nitrates, phosphates) and phytosanitary products (pesticides, pesticides, insecticides).

3. Regulations

Legislations and applications related to groundwater are dictated by the Ground Water Rule (GWR) regulation, explained below:

3.1. Ground Water Rule (GWR)

Title: Ground Water Rule (GWR) 71 FR 65574, November 8, 2006, Vol. 71, No. 216 Correction 71 FR 67427, November 21, 2006, Vol. 71, No. 224

Purpose and General Description

The Groundwater Rule (GWR) is regulated by the EPA, its objective is to provide protection against microbial pathogens or microorganisms in public water systems, as they can be found in fecal contamination and are the cause of many pathogenic diseases. The purpose of the Groundwater Rule is to eliminate the risks associated with drinking water by reducing the occurrence of these.

Applicability of GWR

- Public water systems that use groundwater as a source of drinking water.
- Any system that supplies groundwater and groundwater to consumers where groundwater is untreated

Strategy of GWR

For this, the strategy of the Groundwater Rule (GWR) consists of:
- Sanitary inspections which provide on-site review of how a GWS is maintained and works. It is performed by a trained inspector who reviews water sources of the system, equipment, facilities and treatment procedures. It is done every 3 years. Eight critical elements of a public water system and the identification of significant deficiencies are evaluated. The GWS has 30 days to consult with the state on the appropriate corrective action necessary to address the significant deficiency and has 120 days after the initial status notification to complete the required corrective action and 30 days to notify compliance with the corrective action, otherwise it will become one in a violation of the technique of treatment (TT) for the GWS.

- If the system identifies a positive sample during its water source monitoring the evaluation of the Total Coliform Rule (the State decides) directed at high-risk systems. The Total Coliform Rule (TCR) requires less than 5 samples of TCR per month to receive a sanitary survey every 5 years. If protected and disinfected groundwater is used, the sanitary survey will be every 10 years.
- In case the system has a significant deficiency or fecal contamination of groundwater, these Corrective Action alternatives must be implemented: i) Correct all significant deficiencies, ii) Provide an alternative source of water, iii) Eliminate the source of contamination, iv) Provide treatment which reliably achieves at least 99.99 percent of 4 virus records before the first client.
- Compliance monitoring to ensure that installed technology for drinking water treatment will provide inactivation and/or removal of viruses.

Eight Elements of the Sanitary Survey
- Protection of the Source
- Design, operation, maintenance, and management of the treatment
- Distribution System water quality data
- Finished Water Storage
- Proper working conditions of the Pumps, Pump Facilities, and Controls
- Monitoring, Reporting, and Data Verification
- System Management and Operation
- Operator Compliance with State Requirements

Public Notice to Consumers

A public notice must be sent to consumers if:
- There is a fecal indicator (E. coli, coliphage, enterococci) in its groundwater source (notification within 24 hours).
- Do not collect water samples from the source after a TC-positive sample (notification within one year).
- Do not correct fecal contamination of source water or a significant deficiency within 120 days of notification (notification within 30 days).
- Has any significant deficiency that has not been corrected at the time of your next RCC (or within 12 months for the non-community.
- Failure to comply with performance, monitoring or reporting requirements for any virus treatment (notification within 30 days).

3.2. Revised Total Coliform Rule (RTCR)

Title: Revised Total Coliform Rule (RTCR) 78 FR 10269, February 13, 2013, Vol. 78, No. 30. Before was Total Coliform Rule (TCR) 54 FR 27544-27568, June 29, 1989, Vol. 54, No. 1241.

Purpose and General Description

The Revised Total Coliform Rule (RTCR) and Total Coliform Rule (TCR) have been issued by the Environmental Protection Agency (EPA). The RTCR is the revision of 1989 Total Coliform Rule (TCR) and aims to improve public health protection by eliminating incidents of fecal contamination such as total coliform bacteria: fecal coliforms and Escherichia coli (E. coli) in Groundwater distribution systems.

This law establishes a maximum contaminant level (MCL) based on presence or absence for E. coli and total coliforms. It works with a method of "finding and fixing" fecal contamination. For them, public water systems (PWSs) should have sanitary inspections for systems that collect less than five samples per month for the purpose of identifying and taking corrective actions. The Analytical Methods for Source Water Monitoring for the Fecal Indicator are E. coli, Enterococci and Coliphage.

Applicability of RTCR
The TCR applies to all public water systems (PWSs).

Total Coliforms Rule (TCR)
The Total Coliform Rule (TCR) was published in 1989 and came into force in 1990. This rule determines the maximum contaminant levels (MCL) of total coliforms in drinking water should be zero. EPA determines that PWS should not find coliforms in more than five percent of samples taken each month, otherwise this violation must be reported to the public and reported. If the presence of coliforms is positive, in the next 24 hours a mutation must be collected in at least 5 sampling points.

Monitoring Requirements (TCR)
- The number of routine samples required depends upon system type and the number of people served.
- A written sample siting plan should be review and revised by the State and define the samplings.
- Additional routine samples depend on routine sampling results and count toward calculating MCL compliance.

Main provisions of RTCR are:
- Define a maximum contaminant level target (MCLG) and maximum contaminant level (MCL) for E. coli.
- The treatment of total coliforms is a technical requirement.
- Monitoring according to a sampling plan, location and specific schedule for the E. coli.
- Requirements for seasonal systems to control and certify the completion of state-approved start-up operations.
- In case of contamination: requirements for evaluations and corrective actions
- The public notice (PN) requirements for violations.
- Specific language in the report

Public Notice to Consumers
A public notice must be sent to consumers if:
- Public Notification (PN) is required for violations of the Maximum Contaminant Level (MCL) and monitoring/reporting requirements.
- PWS will be infringing if it does not meet the requirements of RTCR.
- The PWS must use health effects language and inform the public if they do not meet certain RCRT requirements. The type of PN depends on the severity of the violation.
- They must use the specific language when they must make an evaluation or in case of an MCL E. coli noncompliance.

VI.   REMEDIAL ACTIONS

To date, EPA (EPA, 2017) has shown that approximately 157 NPLs (National Priorities List) are contaminated due to hard rock mining activities and are being cleaned by federal agencies. In mining activities, there are many processes that involve the methods of exploitation, the supply of water and the treatment of surface and groundwater. The management of the risk of contamination by acid waters is based on the elimination of these elements that generate this risk: sulfur, oxygen, moisture or catalytic bacteria. The risk of generation of acid waters must be prevented, for them the following steps of Preventive acid water management must be considered:

- Prevent and reduce its generation through the planning of the operations to be carried out, characterizing the possible effluents and effects.
- Detect sources of pollution (equipment or activities) and points of discharge.
- Build barriers.
- Concentrate the effluents to be able to isolate them from the environment.
- Treatment of contaminated water
- Regenerate the medium.
- Quantify effects.
- Heavy metals and leaching
- Chemical Contamination
- Erosion and sedimentation

When the preventive management of acid waters does not work and there is a risk of contamination, water control techniques must be implemented.

Water control techniques

As previously mentioned, the biggest problem in mining is Acid Mine Drainage (AMD), this water is handled with great care to avoid its contamination with the environment. There are water control techniques, which include:

- Interception of the groundwater through the reduction of the water table and the definition of the underground flow, this way it will be known exactly the technical solutions to the problem.
- Divert surface water through the construction of dams upstream to avoid mixing with contaminated water.
- Reuse water for mineral processing, this will help to reduce the volume of water that requires treatment.
- Evaporative removal and evapotranspiration removal: Using water evaporation in ponds, the more it evaporates, the smaller the volume of contaminated water.
- Capture drainage water from the rains at the mine through the installation of coatings, ore piles, covered with rocks, pipes to be able to direct the water to the tailings dams to avoid contact with contaminated waters.
- Reduce the infiltration of rainwater and surface water.

Remedial actions

Remedial actions are defined as long-term cleanings with the aim of providing protection to public health and the environment. To this end, it has designed actions that reduce to zero or minimize the release of pollutants or hazardous substances. In specific response to situations such chemical spill events, state permit requirement, enforcement actions.

There are different methods of remediation with the only goal of cleaning the contaminated water. The effectiveness of treatment methods is around 85-95% of the wastewater pollution. There are three most important methods of remedial actions to eliminate metals and soluble sulfates by immobilization through physical, chemical and biological approaches:

- **Biological treatment technologies**

Iron and sulphate are the major pollutants in the mine's waters. When sulphate minerals dissolve they produce acid waters in the contaminating soils called acid mine drainage (AMD), then these contaminated waters are added neutralization reagents, chemical oxidation of the iron and finally precipitation of sulfate minerals. The metabolism of iron and sulfate is then used through microorganisms, organic matter and plants in order to clean and improve the quality of contaminated water. Treatments include: bio-accumulation, bio-venting and biosparing (when using biological material for the decomposition of wastewater chemical substances), phytoremediation (when plants through roots absorb these pollutants), bioslurping (combining Bio-venting and vacuum pumping to recover products free of groundwater) and permeable reactive barriers.

- **Chemical treatment technologies**

This method considers carbon uptake, ion exchange, chemical precipitation and oxidation in order to have clean water through these chemical reactions. For example, the methods are: chemical precipitation (for wastewater, treatment is used to remove hardness and heavy metals), ion exchange (the water is passed down under pressure through a fixed bed of granular medium), Carbon absorption (in which activated carbon absorbs volatile organic compounds from groundwater by chemically attaching them to carbon atoms), Chemical oxidation (are sent to destroy organic molecules).

- **Physical treatment technologies**

This is the most common method of treatment, including Air sparging (with the help of spray air to remove clean water), Dual phase vacuum extraction (uses a high vacuum system to remove contaminants), Monitoring -Well Oil Skimming (using a belt-style oil skimmer to remove oil and other floating hydrocarbon contaminants) and Pumping-and-Treatment (a more traditional method used for dissolved chemicals like industrial solvents and metals. Soil is treated by biological or chemical means after water is pumped to the surface).

Also, EPA has presented the Best Managerial Practice of the Green Remediation:
- **Characterizing the mining influenced water (MIW):** contaminated surface water, groundwater, and seepage from former mine adits to better understand water mining problems. A characterization of the water resource, the flow of air and water as they are the main actors to weathered the rock causing acid drainage.
- **Using passive treatment systems for acid mine drainage:** utilization of stormwater controls, use the check dams to capture rainwater, use of novel protein-containing food waste to bind metals existing at trace concentrations in water and install remote monitoring equipment for collecting water quality data.

- **Integrating on-site renewable energy to power cleanup operations**: supplement gradient-driven transfer of MIW to or among treatment cells.
- **Installing soil covers:** to stabilize soil and waste piles and reduce their exposure.
- **Reclaiming residual natural resources:** such as economically valuable metals from waste piles. Use of water treatment systems that recover metals from AMD, application of phytotechnologies to treat contaminated water.
- **Integrating cleanup with restoration and reuse:** Promote surface water corridors to complement regional watershed plans.

CHAPTER 3

REPRESENTATIVE STATES REGULATIONS

This chapter lists the most important laws concerning the treatment and quality of water that apply to mining activities in the arid states of the Western U.S. that are key producers of metals and coal.

I. ARIZONA

Prevailing laws and regulations currently governing Arizona's mining and water quality are described in this chapter. Although regulations and programs are created to be applied to all industrial activities, this chapter highlights sections that concern, not exclusively, to mining.

Major pollutant activities in Arizona include agriculture, industrial waste (of any kind), septic tanks, leaking underground storage tanks (LUSTs), landfills, mining facilities and wastewater treatment plants (ADEQ, 2000). Mining activities are responsible for having high levels of dissolved mineral content in certain groundwater areas. Sulfate, Total Dissolved Solids (TDS) and hardness are elevated downstream from historic mining operations and tailings ponds. Heavy metals are naturally present in groundwater, however, elevated levels are associated with mineralized mining districts.

1. Overview
Arizona Revised Statutes (A.R.S.). Title 49. The Environment. Chapter 2. Water Quality Controls

Arizona Administrative Code (A.A.C). Title 18. Chapter 9-11.

2. Definitions

"Administrator" means the administrator of the United States environmental protection agency.

"Aquifer" means a geologic unit that contains sufficient saturated permeable material to yield usable quantities of water to a well or spring.

"CERCLA" means the comprehensive environmental response, compensation, and liability act of 1980, as amended (P.L. 96-510; 94 Stat. 2767; 42 United States Code sections 9601 through 9657), commonly known as "superfund".

"Clean water act" means the federal water pollution control act amendments of 1972 (P.L. 92-500; 86 Stat. 816; 33 United States Code sections 1251 through 1376), as amended.

"Department" means the department of environmental quality.

"Director" means the director of environmental quality or the director's designee.

"Discharge" means any addition of pollutants to navigable waters from aby point source

"Environment" means navigable waters, any other surface waters, groundwater, drinking water supply, land surface or subsurface strata or ambient air, within or bordering on this state.

"Permit" means a written authorization issued by the director or prescribed by this chapter or in a rule adopted under this chapter stating the conditions and restrictions governing a discharge or governing the construction, operation or modification of a facility.

"Pollutant" means fluids, contaminants, toxic wastes, toxic pollutants, dredged spoil, solid waste, substances and chemicals, pesticides, herbicides, fertilizers and other agricultural chemicals, incinerator residue, sewage, garbage, sewage sludge, munitions, petroleum products, chemical wastes, biological materials, radioactive materials, heat, wrecked or discarded equipment, rock, sand, cellar dirt and mining, industrial, municipal and agricultural wastes or any other liquid, solid, gaseous or hazardous substances.

"Standards" means water quality standards, pretreatment standards and toxicity standards established pursuant to this chapter.

3. Water Quality Controls

3.1. Aquifer Permit Required

The Aquifer protection program is regulated under the Arizona Revised Statutes (A.R.S) Title 49, Article 3 and also under the Arizona Administrative Code (A.A.C) Title 18, Chapter 9 (articles 1 through 4) and Chapter 11. Any proposed construction categorized as a discharging facility can be permitted with an Individual Permit to address a specific design or can be covered under a General Permit (typically the case in the mining industry) which authorize a category of discharging within a common geographic area.

The Arizona Department of Environmental Quality (AQEQ) issues the Aquifer Protection Permit (APP) to the subject mining facility–and any other industrial installation–for its entire life, including operation, closure and post-closure periods. The APP document grants the permit and lists the requirements that will allow the permittee to comply with these two key requirements of the Aquifer Protection Program:

- Meeting the Aquifer Water Quality Standards (AWQS) at the Point of compliance (POC)
- Demonstrate the use of Best Available Demonstrate Control Technologies (BADCT), whose purpose is to employ best established engineering controls, processes, operating methods, processes or other techniques, including facilities' site-specific characteristics, in order to reduce discharge of pollutants to the highest degree possible (practicable) before reaching the aquifer or to stop/divert pollutants course to the aquifer. In assessing BADCT, practicable means able to be done from a reasonable standpoint of technical practicality and economically feasible on a mining industry basis, except for substances listed by the Secretary of Health and Human Services (HHS) as being known or reasonably anticipated to be carcinogens or any organic substance listed in 40 CFR section 261.33 (e), from which limit their discharge is practicable to the maximum extent regardless of cost.

Discharging facilities

The Director is the entity responsible for giving public notice that there will be no migration of pollutants to the aquifer and will receive any public comment on the subject. He will make public the name and locations of all facilities that need to acquire an Aquifer Protection Permit.

As defined by the Arizona Revises Statutes (A.R.S), discharging facilities are:

- Any surface impoundment, including disposal pits, storage facilities, ponds and lagoon
- Solid waste disposal facilities (except those that will not be subject to leaching)
- Injection wells.
- Mine tailings piles and ponds.
- Land treatment facilities.
- Facilities that add a pollutant to a salt dome formation, salt bed formation, dry well or underground cave or mine.
- Underground water storage facilities.
- Sewage treatment facilities, including on-site wastewater treatment facilities.
- Mine leaching operations.
- Wetlands designed and constructed to treat municipal and domestic wastewater for underground storage

Facilities Requirements

Application to obtain the Aquifer Protection Permit may be required to furnish the following information:

- Design of the discharge facility and description of its operation which must ensure the greatest degree achievable of discharge

reduction by application of the best available demonstrated control technologies (BADCT)
- Identification and characterization of pollutants discharged in the area and their existing and proposed control measures.
- Definition and characterization of the discharge impact area, including the zone between the surface and water table (vadose zone), through a hydrogeologic study.
- Report of existing quality of aquifer's water and quantity used in the discharge impact area
- Closure strategy
- Permitting fees as annual registration as explained in the following table:

Table 3.1: Annual registration of permittees

Shall register each year based on total daily influent of pollutants	*If capacity > 3,000 gallons per day shall register the permit each year based on total daily discharge of pollutants*
Any surface impoundment	Injection wells
Mine tailings piles and ponds	Land treatment facilities
Facilities that add a pollutant to a salt dome formation or salt bed formation	dry well
	On-site wastewater treatment facilities
Underground cave or mine	
Sewage treatment facilities	
Mine leaching operations	

Depending upon the Director's criteria, it may be prescribed in the permit certain terms and conditions to ensure compliance with the program which includes:
- Regular monitoring
- Reporting requirements and archive
- Establish contingency plans
- Discharge limitations
- Closure, and if the facility requires it, post-closure monitoring and maintenance plans

Relevant information and Director's criteria
As part of the evaluation process to issue the Aquifer Protection Permit, the Director shall consider certain information and criteria relevant to the discharge facility, which includes:
- Estimating quantity and quality (concentration and toxicity) of discharge probable to reach the aquifer from several control technologies and the total cost of the application of technology in relation to the discharge reduction to be achieved by each application
- The extent to which aquifer water quality for beneficial use will be maintained by different control technologies.
- The age of the facility and equipment
- Control processes in place and future changes

Open pit passive containment
A discharging facility that operates in an open pit mine has to satisfy all requirements previously described, and must also meet two conditions: the open pit creates a passive containment that captures the pollutants discharge and that is hydrologically isolated so that there is no discharge of pollution from the capture zone. Such passive containment may be an engineered or natural

control measure that operates without the need for continued maintenance. Additionally, the facility must employ other processes, methods or alternatives to minimize discharge.

Financial competence

Demonstration of financial assurance or competence is required to maintain the permit along its duration until closure or post-closure. Except for state, federal or any local governmental entity, the permittee shall demonstrate financial responsibility and report that the financial mechanisms are kept as scheduled in the permit.

Pollutant management area

If discharging facilities are located in a contiguous geographic area under a common ownership, the application can be submitted collectively for the entire facilities included in the pollutant management area pursuing a single area-wide permit.

Stormwater general permits

For those facilities (catchments, impoundments and sumps) used solely for the management of stormwater and that are governed according to the Clean Water Act, a stormwater general permit is issued if three conditions are met: i) a national pollutant discharge elimination permit has been issued for any stormwater discharges at the facility pursuant to the Clean Water Act or that an application has been submitted for (and not been denied); ii) notification to the director that all requirements set by the first condition have been met; and iii) the owner of the discharge facility has in place a stormwater pollution prevention plan.

Exemptions

Some classes or categories of discharging facilities may be exempted from the Aquifer Protection Permit requirements if in the director's opinion there is no reasonable evidence of degradation of the aquifer water quality, or because the discharging facility already complies with permit requirements of any other federal or state regulations that provide the same or greater aquifer water quality protection. For the purpose of mining interest, exemptions are granted to those excavation sites that receive overburden material that has not been subject to any leaching agent or chemical or process of any kind.

3.2. Arizona Pollutant Discharge Elimination System Program (AZPDES)

AZPDES is referred to the Arizona pollutant discharge elimination system program as adopted under section 402(b) of the Clean Water Act, which establishes a comprehensive program "to restore and maintain the chemical, physical, and biological integrity of the US waters" (33 U.S.C. § 1251a). In order to attain this purpose, the Clean Water Act requires Environmental Protection Agency (EPA) to authorize discharges via issuance of National Pollutant Discharge Elimination System (NPDES) permits. On the first phase of the NPDES, the EPA publishes a regulation on "stormwater discharges associated with industrial activities" on November 16, 1990 (55 FR 47990). The Arizona Department of Environmental Quality (ADEQ) received the authorization to administer the program in the state of Arizona, except for Indian Country, on December 5, 2002.

3.3. Enforcement

If a person is found in violation of an adopted rule or permit issued condition that is creating imminent or substantial damage to the environment and public health, the director may issue an order of compliance, specifying the nature of the violation, and a reasonable time for compliance. In observing the law, each day of violation constitutes a separate violation. It is unlawful:
- Discharge pollutants without an issued permit
- Fail to observe conditions on an issued permit as to monitor, sample or report discharges
- Violate discharge limitations and water quality standards
- A person that knowingly or recklessly and/or criminal negligence performs an act prohibited by rules and regulations
- A person who manifests an extreme indifference for human life in conducting a prohibited act

Table 3.2. Penalties, fines and/or remedies – Arizona

Violation	Penalties, fines and/or remedies
Violation of a rule, permit, order of discharge limitation	Civil penalty, not more than $25,000/day of violation.
Alter or modify any monitoring device or method;	Class 5 felony
Fail to maintained, operate or repair the monitoring device;	Class 5 felony
Discharge without a permit	Class 5 felony

II. COLORADO

The Water Quality Control Commission is responsible for the creation and implementation of water quality policies in the state of Colorado, which are established by the Colorado Water Quality Control Act. To this end, the commission has classified and established water quality standards to protect the use of Colorado waters. This classification and norms are made by means of permissions for the discharges of point sources in the majority. It aims to create state standards, classification systems for surface water and set standards in relation to groundwater and surface water.

The Water Quality Control Commission function includes Regulatory hearings, Commission approval administrative action, Informational Hearings, Adjudicatory Hearings and taking formal action in policy statements.

However, the main agency responsible for this implementation of discharge permit programs in Colorado is the Water Quality Control Division.

1. Overview
Colorado Water Control Act (Colo. Rev. Stat. § 25-8-100, et seq.), the Colorado Code of Regulations (5 CCR 1002-21, et seq.)

2. Definitions

"Discharge of pollutants" means the introduction or addition of a pollutant into state waters.

"Division" means the division of administration of the Department of Public Health and Environment.

"Issue" or *"issuance"* means the mailing to all parties of any order, permit, determination, or notice, other than notice by publication, by certified mail to the last address furnished to the agency by the person subject thereto or personal service on such person, and the date of issuance of such order, permit, determination, or notice shall be the date of such mailing or service or such later date as is stated in the order, permit, determination, or notice.

"Person" means an individual, corporation, partnership, association, state or political subdivision thereof, federal agency, state agency, municipality, commission, or interstate body.

"Point source" means any discernible, confined, and discrete conveyance, including, but not limited to, any pipe, ditch, channel, tunnel, conduit, well, discrete fissure, container, rolling stock, concentrated animal feeding operation, or vessel or other floating craft, from which pollutants are or may be discharged. "Point source" does not include irrigation return flow.

"Pollutant" means dredged spoil, dirt, slurry, solid waste, incinerator residue, sewage, sewage sludge, garbage, trash, chemical waste, biological nutrient, biological material, radioactive material, heat, wrecked or discarded equipment, rock, sand, or any industrial, municipal, or agricultural waste.

"Pollution" means the man-made, man-induced, or natural alteration of the physical, chemical, biological, and radiological integrity of water.

"State waters" means any and all surface and subsurface waters which are contained in or flow in or through this state, but does not include waters in sewage systems, waters in treatment works of disposal systems, waters in potable water distribution systems, and all water withdrawn for use until use and treatment have been completed.

"Water quality standard" means any standard promulgated pursuant to section 25-8-204.

3. Water Quality Controls

3.1. General Provisions

The State of Colorado's legislative declaration is profoundly sound in its objectives that is worth quoting in entirely (From article 25-8-102. Legislative declaration): *"In order to foster the health, welfare, and safety of the inhabitants of the state of Colorado and to facilitate the enjoyment and use of the scenic and natural resources of the state, it is declared to be the policy of this state to prevent injury to beneficial uses made of state waters, to maximize the beneficial uses of water, and to develop waters to which Colorado and its citizens are entitled and, within this context, to achieve the maximum practical degree of water quality in the waters of the state consistent with the welfare of the state. It is further declared that pollution of state waters may constitute a menace to public health and welfare, may create public nuisances, may be harmful to wildlife and aquatic life, and may impair beneficial uses of state waters and that the problem of water pollution in this state is closely related to the problem of water pollution in adjoining states"*

3.2. Water Quality Control Commission

A commission to exercise state power and perform duties as they were transferred from the department of public health and environment. The commission will consist of nine citizens of the state members, appointed by the governor, with the Senate consent for terms of three years each.

Members of the commission can be removed by the Governor for the following reasons: failure to assist to two consecutive meetings without a valid reason, malfeasance in office or any cause that proves the member incapacity for the role.

The commission shall point a chairman, vice-chairman and secretary, selected under the observation of public meetings and members quorum (two-thirds of the member's number constitute a quorum).

Classification of state waters

Waters of the Colorado state are classified based upon relevant characteristics:
- Existing and maximum extent of pollution and sources types
- Beneficial uses of water
- Type and uses of land surrounding the waters
- Need to improve water quality
- Classify waters on type (such as surface or groundwater), lakes or streams, depth, volume, temperature, variability and other characteristics.

Assigned classifications with regards to major river basin location:

Table 3.3. Classification in regards to river basin location

Regulation 31	Basic Standards and Methodologies for Surface Water
Regulation 32	Classifications and Numeric Standards for Arkansas River Basin
Regulation 33	Classifications and Numeric Standards for Upper Colorado River Basin and North Platte River
Regulation 34	Classifications and Numeric Standards for San Juan River and Dolores River Basins
Regulation 35	Classifications and Numeric Standards for Gunnison and Lower Dolores River Basins
Regulation 36	Classifications and Numeric Standards for Rio Grande Basin
Regulation 37	Classifications and Numeric Standards for Lower Colorado River Basin
Regulation 38	Classifications and Numeric Standards for South Platte River Basin, Laramie River Basin, Republican River Basin, Smoky Hill River Basin
Regulation 39	Colorado River Salinity Standards

Water quality standards

Promulgated by the commission with respect to any measurable water parameter applicable to a designated area of the state or to all state waters, including, but not limited to:
- pH and hydrogen compounds
- Chlorine and heavy metals
- Toxic substances
- Suspended solids and colloids

- Bacteria, fungi, viruses, fecal coliform and others biological beings
- Dissolved oxygen
- Phosphates and nitrates
- Salinity, alkalinity, and acidity
- Temperature
- Taste, color, odor, and turbidity
- Oil and grease, trash and refuse

Control regulations

The commission may promulgate prohibition, standards, and any other limitation for the following purposes:
- Specifically identify pollutants
- Wastes effluents
- Sewage treatment works
- Use of disposal biosolids
- Reuse of reclaimed domestic wastewaters and greywater

Water quality designations

Definitions and designation of water by the commission, on either outstanding or use-protected water:
- Outstanding waters, designated to waters that shall be maintained at the existing quality.
- Use-protected waters, meaning waters whose quality is no better than necessary to support propagation of aquatic and wildlife
- Reviewable waters, for those waters not designated as outstanding or use-protected.

3.3. Procedures

The commission shall hold public hearings, ruled and regulated by the commission itself, by adopting what is necessary to make sure it is fair and impartial. Any person can propose a new regulation or a change in factual regulation by filing a written proposal to the commission not less than twenty-days prior the public hearing. The proposal shall be in the public domain.

Judicial review

Any final determination by the commission shall be subject to a judicial review, including but limited to permits, regulations, orders (enforcement, cease, and clean-up), water quality standards, site approvals. Any proceedings shall be filed in the local court where pollution takes place.

3.4. Permit System

Discharge of pollutants

Without first having a permit, any person shall discharge any pollutant into the waters of the state from any point source. Nor shall discharge pollutants into ditches or man-made conveyance for the purpose of evading the permit requirement. Issuing a discharge permit requires applications to have the following content, among other matters:

- Identification of the applicant and facilities
- Quantity and quality of the discharge
- Treatment conditions plans prior discharge
- Reporting procedures and methods used for monitoring
- Schedule of compliance
- Description of facility functionality and plans
- Contingency plans for the facility
- Duration of the permit requested

- Notice of intent

Effective Permit Fess

The commerce and industrial sector require an annual fee schedule for regulated activities related to mining, sugar processing, industrial stormwater, hydrocarbon refining, and others:

Table 3.4. Annual fees

	Annual fees
Sand, gravel and placer mining	
a) Pit dewatering only	$500
b) Pit dewatering and wash water discharge	$570
c) Mercury use and discharge impact	$640
d) Storm water discharge	$435
Coal mining	
a) Sedimentation ponds, surface runoff	$980
b) Mine Water, preparation plant discharge	$1,320
Hard rock mining	
a) Mine dewatering from 0 to 49,999 gallons/day	$1,140
b) Mine dewatering from 50k to 999,999 gallons/day	$2,150
c) Mine dewatering for greater than 1,000,000 gallons/day	$3,280
d) Mine dewatering and milling – no discharge	$3,280
e) Mine dewatering and milling – with discharge	$9,880
f) No discharge	$1,140
g) Milling and discharge from 0 to 49,999 gallons/day	$3,350
h) Milling and discharge 50,000 gallons/day and more	$6,680

General Permits

a) Sand, gravel and process discharge and storm water	$270
b) Sand, gravel and storm water	$75
c) Placer mining	$520
d) Coal mining	$780
e) Mines less than ten acres – storm water only	$125
f) Mines ten acres and more – storm water only	$375
g) Inactive mineral mines – storm water only	$75
h) Coal degasification – process 0 to 49,999 gallons/day	$2,150
i) Coal degasification – process 50,000 to 99,999 gallons/day	$3,280
j) Coal degasification – process 100,000 gallons/day or more	$9,880

For operating industrial operations and water treatment plants that discharge into a domestic wastewater treatment, the following fees apply:

Table 3.5. Fees for water treatment plants

	Annual fees
Water treatment plants	
a) Intermittent discharge	$500
b) Routine discharge	$570
Coal mining	$640
a) Water treatment plant, intermittent discharging	$435
b) Water treatment plant, routine discharging	
c) Discharge for a population of 3,000 or less	$980
d) Discharge for a population of 3,301 to 9,999	$1,320

e) Discharge for a population of more than 10,000	$1,140

3.5. Penalties

Any person can apply to the divisions to report any accidental discharge or suspected violation to any rule, regulation, permit or order promulgated. For spills or discharges that can cause pollution to waters of the state, the person engaged in those operations shall notify the accident to the division, as soon as the person is knowledgeable.

Notice of alleged violations

A written notice of alleged violation shall be caused to a person suspected of an alleged violation to a permit, regulation or order promulgated by the state authorities. Notice may include any corrective action required. The suspected violator shall require to answer each violation and may require appearing in a public hearing to answer those alleged violations, in a time no sooner than fifteen days after service of the notice.

If the division finds that a violation of any order, regulation or permit exists, it may issue a cease-and-desist or clean-up order, and the appropriate legal actions.

Table 3.6. Penalties, fines and/or remedies - Colorado

Violation	Penalties, fines and/or remedies
Fail to notify the division of a spill or discharge of oil or substance that can cause pollution to the state waters	Guilty of misdemeanor. Penalty fine for not more than $10,000 or/and prison for not more than 1 year
Violate a permit issued, control regulation of a final order	Penalty fine not more than $10,000/day the violation occurred.
Criminal pollution (intentional or with criminal negligence) discharging pollutants to state waters, violating permits or orders, or without having obtained a permit	Criminal negligence: Penalty fine for not more than $12,500. Intentionally: Penalty fine of not more than $25,000. If a second of these violations occurred in a period of fewer than 2 years, fines amounts shall be doubled
Falsification and tampering	Guilty of misdemeanor. Penalty fine for not more than $10,000 or/and prison for not more than 6 months

In determining the penalty amounts, it shall be considered:
- Damage from the violation
- Legal history of the violator
- Intentionally of the violation

- Duration of the violation
- Violator economic benefit realized by the violation

Penalties collected shall be transferred to the state treasurer, who shall credit all amounts to the water quality improvement fund.

III. NEVADA

Nevada is a world-leading producer of gold (two/thirds of American gold production), silver, copper, lithium and a long list of other minerals. The Bureau of Mining Regulation and Reclamation (BMRR) is the entity responsible for administering the Nevada's environmental regulations on mining through three of its branches: regulation, closure, and reclamation within the Nevada Division of Environmental Protection (NDEP).

Nevada is divided into hydrological basin zones, and most of its water consumption comes from underground pumping. Although in certain zones such Las Vegas or Reno, water is supplied from surface water sources of the Truckee and Colorado rivers. Mining, milling, and beneficiation facilities are required to submit an application to obtain a water pollution control permit before any construction begins. Permits are designed to protect the waters of the state from degradation and mine operations using chemicals in their processes, as it is the case of gold, are required to meet a zero discharge policy and to comply with the Clean Water Act and Safe Drinking Water Act (NRS Chapter 445A).

The process of issuing permits to mining operations, and in general to any industrial facility, is accessible to the public through media circulation and hearings prior to the final notice of approval or denial. Extensive research, discussion, and opinion exchanges have taken please between interested parties. The U.S. Geological Survey, non-profit organizations, state authorities and other actors have been altogether monitoring and auditing mining companies, especially Newmont gold and Barrick Goldstrike, the largest gold mine producers in Nevada. A long-term concern is the significant dropped levels of the water table in parts of Nevada where mine facilities are actively pumping millions of gallons of water out of the groundwater, resulting in a water table falling off as much as 1,000 feet around the largest open pit mines, though the average waterfall is about 70 feet since the 60s (USGS, 2000). Now, pumped water is being poured onto the rivers, increasing the flows of the Humboldt River, but what environmentalists fear is when mining activities, so pumping, come to an end. The river could dry up since water flows would infiltrate back to the emptied groundwater deposits. However, a possible solution to include in the closure and reclamation plans may be the creation of pit lakes, through which water can be distributed to farmers after mining ceased.

1. Overview

Permit system is supervised by the Office of State Engineers (OSE)

Statutes are set forth in Chapters 533 and 534 of the N.R.S. Nevada Revised Statutes

Nevada Administrative Code -445A Revised Date: 6-16.

U.S. Army Corps of Engineers under Section 404 of the Clean Water Act.

2. Definitions

"Area of review" means the area surrounding a facility which is to be evaluated

"As-built drawings" means engineering drawings which reflect all changes made from original engineering drawings during the construction of a facility so that a representation of the facility as constructed is portrayed.

"Beneficiation" means the dressing or processing of ores for:

1. Regulating the size of a desired product;
2. Removing unwanted constituents; and
3. Improving the quality, purity or assay grade of a desired product.

"Best engineering judgment" means that decision by the Department which, after evaluating the available alternatives and levels of technology presented by the applicant, results in an acceptable design for containing contaminants from a facility in order to protect the waters of the State.

"Commission" means the State Environmental Commission.

"Department" means the State Department of Conservation and Natural Resources.

"Facility" means all portions of a mining operation, including, but not limited to, the mine, waste rock piles, or piles, beneficiation process components, processed ore disposal sites, and all associated buildings and structures. The term does not include any process component or non-process component which is not used for mining or mineral production, and has not been used in the past for mining or mineral production as part of an operation which is active as of September 1, 1989.

"Groundwater" means all subsurface water comprising the zone of saturation, including perched zones of saturation, which could produce usable water.

"Liner" means a continuous layer of man-made or reconstructed natural materials, or a combination thereof

which restricts the downward or lateral movement of liquids.

"Mining" means the process of extracting ores from the earth.

"Ore" means the naturally occurring material from which a metallic mineral of economic value can be extracted.

"Permit" means a written document issued pursuant to NRS 445A.300 to 445A.730, inclusive, which describes the responsibilities and obligations of the holder of the permit during the construction, operation, and temporary or permanent closure of a facility.

"Pond" means a process component which stores, confines or otherwise significantly impedes the horizontal movement of process fluids. The term does not include tailings impoundments, vats, tanks or other non-earthen containers.

"Source" means any building, structure, facility or installation from which there is or may be the discharge of pollutants.

"Storm event" means a precipitation event with a specified frequency of return and specified period of duration as defined in Precipitation-Frequency Atlas of the Western United States, vol. VII-Nevada.

"Tailings impoundment" means a process component which is the final depository for processed ore discharged from a mill.

"*Zero discharge*" means the standard of performance for the protection of surface waters which requires the containment of all process fluids.

3. Water Quality Controls

3.1. Permit for Facilities

Almost all mining operations in Nevada shall submit an application permit at least 165 days prior to facility construction plans. The procedure includes:
- Before submitting the permit application, it is required to set up a meeting with a representative of the Department to discuss information concerning the mining facility in terms of proposed location, design and physical characteristics of the process component and production plans.
- Application should clearly define site conditions, processing materials and description of probable impacts on the area. Failure to provide this information may result in Department requiring higher engineered containment standards and/or monitoring.
- Relevant information that the permit must contain includes an assessment of the area, a meteorological report, and engineering design, a proposed operating plan and a report on lab sample analysis.

Assessment of the area
A detailed assessment of the area of review must include: i) A hydrogeological and lithological study which defines the surface and subsurface conditions of the facility site adjacent and beneath all point sources to a minimum depth of 100 feet; ii) accompanied

with a geological map of the area within a 1 mile radius of the facility; iii) Topography map for the subject area, identifying all known surface waterways, streams and seeps and existing habitable buildings with 1 miles radius distance; iv) Up-gradient watershed boundaries and potential effects of a 100-year, 24-hour storm event on the facility and v) map showing all wells constructed for supply of drinking water within 5 miles down-gradient of the process components.

Meteorological report
Content of application must prepare a complete overview of rainfall conditions, reporting: i) Historical mean monthly rainfall obtained from the nearest recording station of the mine site; ii) Rainfall events (24-hours) with intervals of recurrence of 10, 25 and 100 years; iii) Historical temperature variations and iv) Sample results from the mine that represents the overburden material, waste rock and ore, characterized by a multi-element spectrographic assay or equivalent procedure and evaluation of potentially pollutants.

Engineering design report
An engineering design report prepared and submitted by a professional engineer registered in the Nevada Board of Professional Engineers and Land Surveyors, which must include:
- Engineering plans for all components of the processing facility used for beneficiation and their general specifications
- Topography and maps showing the mine site, process components for beneficiation, waste rock dumps locations, run-of-mine disposals.
- Maps indicating design of the structures built to control and manage the process fluids

- Methods for controlling the stormwater runoff
- Description of geological and hydrological conditions adjacent and beneath the facilities that will provide natural containment and structural stability
- Description of material for the liner and subbase preparation, construction procedures for leach pads, reservoirs, ponds and ditches.
- Detailed plans for leak detection and adequate monitoring systems.

Proposed operating plans

For the mineral processing circuits, the applicant must submit detailed information that includes descriptions of the processing circuits as well as flow charts indicating the range of operating condition for which the process components were designed. A plan to manage the process fluids must be prepared, which is able to quantify fluids at any moment and to determine what extent the containment has been exceeded because of meteoric water. The monitoring plan will describe: i) quality of water; ii) locations and frequency for monitoring in order to evaluate the groundwater and surface water that may be affected by the subject facilities; iii) location of the leak detection system, frequency of sampling and analytical profile method. Additionally, a response emergency plan (actions and actors) to proceed in case of failures in the fluid management system that can result in the release of pollutants. It is also required to present a tentative temporary and permanent closure plan, indicating the procedures, methods, and schedule to establish spent process materials.

Low-temperature mine sites

Contingency plans due to weather conditions are considered in Nevada legislation. If a facility is located in an area where the average temperature can be as low as zero degrees centigrade for 30 days or more, a plan for a seasonal closure must be prepared as part of the application. The plan has to address the impacts of the closure, identify what activities will be shut down and which ones will be continued, and elaborate a plan to reopen the facility.

Table 3.7: Procedures durations

Application stage	length	remarks
Submitting application	165 days	prior to initiating facility construction
Written notification from the department	30 days	within 30 days after receiving the application (procedurally complete)
Correct missing information	365 days	to submit required information not provided by applicant
Draft permit issued or Notice of intent to deny the permit (Public notice)	90 days	within 90 days after determining that application is technically complete
Make publish notice of intent	30 days	before issuing draft permit or intent of denial

Request a public hearing *(from any interested person)*	30 days	within 30 days before issuing draft permit or intent of denial
Renewal of permit	5 years	If apply for renewal

Pilot or testing facilities

A permit may be issued to construct, operate and permanent close a pilot or testing facilities that will not evaluate more than 100,000 tons of ore and may not exceed 1 year for a single test or 2 years for several tests. Such facility must clearly prove that will not degrade the water quality of the zone.

Modifications on existing permits

Table 3.8: Modifications on permits

Minor Modification	*Major modification*
- doesn't require public notice	- requires public notice
- may not extend the term of the permit	- may extend the term of permit for not more than 5 years
- filled for phased expansion of milling or tailings impoundment or the leach pad, but using the same or equivalent technologies	- include addition of beneficiation process (heap leaching, milling) or significant change of location of process components or change on the proposed beneficiation process

Minimum design criteria

Provisions on the criteria for the design of each process components are based on site characterization (operating conditions), and best engineering judgment to determine the extent to which designs will provide water protection through engineered containment. The following applies to all process components:

- If the subject area annual level of evaporation is greater than precipitation, the facility must achieve zero discharge.
- All process components must be designed so that any release from them will not degrade the waters of the state.
- All components must be designed to withstand the runoff of a 24 hour, 100-year storm event
- Fully functionality must be guaranteed (5 years) for the primary fluid management system design

Leach pads and Non-impoundment surfaces

Certain requirements must be met when designing leach pads and non-impoundment surface to contain and promote the horizontal flow of fluids:

- Solutions must exert minimal hydraulic head on the liner
- Containment design of fluids must consist of a liner system that provides at least an equal containment than that provided by a synthetic liner installed on top of a pre-liner subbase of ½ inches native or soil material, which a maximum coefficient of permeability of 1×10^{-6} cm/sec; or 1×10^{-5} cm/sec if combined with a leaking detection system
- At the discretion of the Department, all open channels may require the installation of a leaking detection system

Ponds

All ponds intended to contain fluids shall have a primary synthetic liner, followed by a layer of a material with the ability to rapidly transport fluids to a collection point, and finally covered by a secondary liner. The material between the liners shall guarantee the capture, transportation, and removal of all fluids so that it prevents the transference of a hydraulic head from the primary liner to the secondary liner, otherwise the pond must be shut down.

For ponds built with the sole purpose of containment excess quantities of process fluids from storm events for limited periods, the Department may approve the construction with one primary liner.

For ponds containing non-process fluids, no liner may be mandated if approved by the Department.

Vats and tanks

For vats and tanks that contain process fluids and can be visually inspected, there is no need of double lining if an area for secondary containment capable of holding 110 percent of the largest container is provided.

Tailings impoundments

The following conditions are applied to the containment system of tailings impoundments:

- Containment must be twelve inches of compacted native or soils with a coefficient of permeability of no more than 1×10^{-6} cm/sec, or

- Having the site underlined by a competent bedrock that has demonstrated to provide containment equivalent to the previous condition

Liners

The following design parameters must be observed:
- Soil liners, if graded on native materials, must have a 12 inches minimum thickness and be placed in lifts of not more than 6 inches thick. Additionally, soil liners must be of not more than 12 inches and have a coefficient of permeability of 1×10^{-7} cm/sec, except for tailings impoundment liners
- Synthetic liners must have a coefficient of permeability of 1×10^{-11} cm/sec
- Overall type of foundation and slope stability
- Provision for hydraulic relief on the over liner
- Compatibility between the process solution and liners
- Proximity of surface to all groundwater
- Liner's capability of being competent until facilities closure

3.2. Discharge Permit

Any person shall discharge any pollutant from a source point to the state waters without having obtained first a permit from the Department, except as otherwise provided.

An initial permit application shall be submitted at not less than 180 days in advance of the discharge. After receiving notification from the Director, the person owner of the facility will have 90 days to submit a complete application, including the following signatures:

- The Principal Executive Officer of the company, or at least the vice president, or the authority responsible for the overall operation
- A general partner, if any, of the facility
- The proprietor
- Authority from the municipal, state or other public facilities (in the rank of Principal Executive Officer)

Fees

Application's fees are non-refundable, and must accompany every original permit application for discharge, as follows:

Table 3.9. Discharge fees

Type of discharge	Fee for original application	Fee for renewal application	Annual fee
Discharge to Groundwater from the Dewatering of a mine			
a) Cooling water only	$625	$315	$1,000
b) < 50,000 gallons/day	625	315	1,500
c) 50,000 to 1,000,000 gallons/day	875	440	2,000
d) 1,000,000 to 5,000,000 gallons/day	1,000	500	2,500
e) >5,000,000/day	1,250	625	3,000

Type of discharge	Fee for original application	Fee for renewal application	Annual fee
Mining			
a) Discharge from Physical separation facility, no chemicals added	$500	$500	$250
b) Mine facility that chemically process < 18,250 tons/day	500	500	250
c) Mine facility that chemically process 18,250 to 36,500 tons/day	1,500	1,500	2,000
d) Mine facility that chemically process 36,500 to 100,000 tons/day	4,000	4,000	4,000
e) Mine facility that chemically process 100,000 to 500,000 tons/day	6,000	6,000	8,000
f) Mine facility that chemically process 500,000 to 1,000,000 tons/day	10,000	10,000	10,000
g) Mine facility that chemically process 1,000,000 to 2,000,000 tons/day	14,000	14,000	14,000
h) Mine facility that chemically process > 2,000,000 tons/day	20,000	20,000	20,000

i) Closed facilities - Monitoring	250	250	500

Type of discharge	Fee for original application	Fee for renewal application	Annual fee
Runoff of Storm Water			
a) Industrial facility size between 5 to 10 acres	$300	$150	$750
b) Industrial facility size between 5 to 10 acres	600	300	750
c) Industrial facility size between 5 to 10 acres	1,000	500	750
d) Municipality with 250,000 population or less	600	300	750
e) Municipality with > 250,000 population	1,000	500	1,000

3.3. Enforcement

Enforcement of any provision of these regulations are served by the Attorney General at the request of the director and the commission.

Civil penalties

Any person who violates any provision of the statutes, regulations, permits, standards or final orders issued (except for provisions concerning a diffuse source), shall pay a fine of not more than $25,000 /day of the violation duration.

Criminal penalties

Any person who intentionally, or acting with criminal negligence violates any statute, regulation, established limitation, permit issued, or any final order (except a final order for diffuse sources) is guilty of a misdemeanor and shall pay a fine of not more than $25,000 /day of the violation duration, or shall be punished by imprisonment for not more than 1 year, or both pay the fine and imprisonment.

Falsification or tampering

A person who makes any false report, plan, certification, representation, statement, application or other document files or required to be monitored and maintained is guilty of a gross misdemeanor and shall pay a fine of not more than $10,000 or by imprisonment or both.

IV. WYOMING

The Wyoming constitution (Article 8, Section 1) declares that all water within the state boundaries is the property of the state. The Wyoming Department of Environmental Quality (WDEQ) administers the general environmental protection rules and regulations, and the Wyoming State engineer's Office oversees all water appropriations. Priority of appropriation rights can be issued to anyone who plans to make a beneficial use of the water, including mining.

Wyoming coal mining is the most prolific in the nation, and all its coal mines combine a 41 percent of the U.S. coal production, followed from afar by West Virginia with 11% percent (DOE-EIA, coal data 2017). Coal mining producers in Wyoming also have a great level of technical sophistication, reaching very efficient recovery factors of 92 percent, and representing an important source of tax revenue for the state and local government, contributing over $1 billion in annual revenue (Wyoming State Geological Survey, 2016).

Wyoming law on water requires permit filing for use and discharge of surface and groundwater through general and individual permits, making sure the potential effects caused by the mining facility will be assessed on groundwater levels and quality by frequent monitoring water samples taken from strategic points, as observation wells and discharge points. Water pumped from the open pits is collected in ponds and treated before being released. Effluents and airborne effluents are monitored periodically to ensure that the mine operation and the mill do not exceed acceptable limits. The WDEQ has developed specific requirements for coal mine operations, in a set of laws apart from other mining metals, in all environmental aspects, not only water. The result is perhaps a very demanding regulation, as it is viewed among adults in Wyoming, according to a poll carried out at every state, indicating that 55 percent of adults in Wyoming think that laws are strict and hurt the state economy (Pew Research Center, 2015).

1. Overview

Relevant Rules and Regulations concerning mine operations and water:

Wyoming Department of Environmental Quality (WDEQ), water quality program:

Chapter 2, Permit Regulations for Discharges to Wyoming Surface Waters

Chapter 3, Permit to construct

Chapter 4, Releases of Oil and Hazardous Substances into Waters of the State

Chapter 8, Quality Standards for Wyoming Groundwater

Chapter 9, Groundwater Pollution Control Permit

Chapter 24 Class VI Injection Wells and Facilities (Carbon Sequestration)

Chapter 27 Underground Injection Control Program (Class I and Class V Injection Wells)

Wyoming State Engineer's office, ground water and surface water program:

Chapter 2: Obtaining a Ground Water Right

Chapter 1: Procedure for Obtaining Surface Water Right

2. Definitions

"Administrator" means the administrator of the Water Quality Division, Wyoming Department of Environmental Quality.

"Ecological function" means the ability of an area to support vegetation and fish and wildlife populations, recharge aquifers, stabilize base flows, attenuate flooding, trap sediment and remove or transform nutrients and other pollutants.

"Nonpoint source" means any source of pollution other than a point source.

"Point source" means any discernible, confined and discrete conveyance, including but not limited to any pipe, ditch, channel, tunnel, conduit, well, discrete fissure, container, rolling stock, concentrated animal feeding operation or vessel or other floating craft, from which pollutants are or may be discharged.

"Pollution" means contamination or other alteration of the physical, chemical or biological properties of any waters of the state, including change in temperature, taste, color, turbidity or odor of the waters or any discharge of any acid or toxic material, chemical or chemical compound, whether it be liquid, gaseous, solid, radioactive or other substance, including wastes, into any waters of the state which creates a nuisance or renders any waters harmful, detrimental or injurious to public health, safety or welfare, to domestic, commercial, industrial, agricultural, recreational or other

legitimate beneficial uses, or to livestock, wildlife or aquatic life, or which degrades the water for its intended use, or adversely affects the environment. This term does not mean water, gas or other material which is injected into a well to facilitate production of oil, or gas or water, derived in association with oil or gas production and disposed of in a well, if the well-used either to facilitate production or for disposal purposes is approved by authority of the state, and if the state determines that such injection or disposal well will not result in the degradation of ground or surface or water resources.

"Best Management Practices (BMPs)" means schedules of activities, prohibitions of practices, maintenance procedures, and/or other management practices to prevent or reduce the pollution of "waters of the state." BMPs also include treatment requirements, operating procedures, and practices to control plant site runoff, spillage or leaks, sludge or waste disposal, or drainage from raw material storage

"Outfall" means the point at which a discharge exits the final treatment unit, if any, associated with a facility prior to entering surface waters of the state.

"Point of compliance" means a point downstream from the outfall where effluent limitations specified in a permit must be achieved.

"Storm water" means storm water runoff, snow melt runoff, and surface runoff and drainage.

"Wyoming Pollution Discharge Elimination System (WYPDES)" means the state program for issuing,

modifying and reissuing, terminating, monitoring and enforcing permits for discharging pollutants into surface waters of the state under the provisions of these rules, W.S. 35-11- 101 through 35-11-1803 and the CWA.

3. Water Quality Controls

3.1. Permits for discharges to surface waters

Regulations are dictated in conformity with requirements stated for Environmental Quality Act and the National Pollutant Discharge Elimination System (NPDES) in order to institute a permit issuance program for point source discharge into surface the of the state.

General Permits
Issued to any categories of discharges, though additional requirements are mandated to effluents permits, stormwater and isolated wetlands. Exceptions are individual permits otherwise covered. General permits regulate:

- Stormwater point sources except for those associated with industrial activities (mining for instance) that can potentially reach surface waters
- Point sources discharges into isolated wetlands
- Effluent discharges

Any person seeking authorization to discharge shall submit a notice of intent, providing information for adequate program implementation. A complete determination granting the discharge permit takes 30 days after submitting the notice of intent. General permits are issued for a specific duration on a case-by-case basis, which may not exceed 5 years.

Individual Permits

The administrator may require the person seeking for a general permit to apply for an individual permit in case:
- There is no compliance with the WYPDES regulations
- Change in demonstrated technologies applicable to the subject point source
- There is a significant contribution of pollution from the discharge (location, size, quantity of pollutants discharged and any other relevant factor)

Effluent Permits

Directed to new facilities, modification on actual permits and renewals of effluent permits:
- New facilities shall submit the application 180 days in advance prior the intended discharge data, providing detailed information about the company or individual seeking the permit, a complete description of the discharge activities and explanation of treatment facilities used supported with topographic maps, duration and quantity (flow rates) and quality of the effluent.
- Public participation on meeting is contained in this regulation
- Issued permits shall comply with the Clean Water Act (Section 307a) standards and prohibition set for toxic pollutants.

Stormwater Discharges

The permit application shall be submitted 60 days in advanced the storm water discharge if the administrator determines it could contribute to any significant law violation or to a major

discharge of pollutants to the waters of the state. A complete application for an individual permit is required, containing the following information, among others:
- Detailed information about the applicant and description if activities being conducted that needs to obtain the WYPDES permit, including topographic maps with the location of drainage and discharge structures, buildings, areas where herbicides, pesticides, soil conditioners, and fertilizers were applied.
- Tests certification of all outfalls that contains stormwater discharges in the area
- Quantities data of pollutants presence, oil and grease, pH, flow measurements from stormwater events sampling

Isolated Wetlands

For applications seeking implementation of mitigation plans caused by activities that endanger naturally occurred isolated wetlands (such activities includes the construction of artificially isolated wetlands used to mitigate the loss of naturally occurred isolated wetlands), from which applies:
- If the activity causes the loss of at least one (1) acre of wetland habitat, submission of a notice of intent for a general permit is required, including a mitigation plan to offset the loss of wetlands. Mining activities are exempt from this requirement if subjected to a permit or authorization by the WDEQ, Land division.

Requirements for Coal Mining operations

Coal mining operations are required to provide additional specific information to discharges, as follows:

- Construction requirements, including a sedimentation structure control plan for ponds or runoff facilities operating in a surface coal mine. Plans must outline a planned scope, inlet ditches designs (inlet ditches shall minimize erosion, disturbance of the pond bottom and resuspension of silts or colloidal soil particles), outlet structures (if used, shall minimize floating solids from discharging without eroding or disturbing the dike)
- Calculations and drawings showing the volume of runoff from a 10 year, 24 hours rainfall event. The design for the runoff facility must include all supporting documentation.
- If pond bottoms and sidewalls are built from fill material, the soil shall have low permeability, be free of trash and organic material and be relatively incompressible.
- Overall slopes for outer dike shall not be steeper than 1:2 horizontal to vertical ratio.
- Dewatering devices for draining the pond for storage resulting from inflow
- Effluent limitations must be met for any point sources mine discharge that enters surface water (except for rainfall events greater than 10-years, 24-hours)

Table 3.10. Effluent limitations

Effluent parameter	30 day average	Maximum daily	Instant maximum
Total suspended solids (mg/l)	35	70	90
Total Iron (mg/l)	3.0	6.0	9.0
Total Manganese (mg/l)	2.0	4.0	6.0
pH (standard units)	N/A	N/A	6.0 to 9.0

- Reclamation areas shall obey the following discharge limitations:

Table 3.11. Limitation during rehabilitation

Pollutant characteristics	Limitations
Settleable Solids	0.5 ml/l maximum
pH (standard units)	Within 6.0 and 9.0 at all times

3.2. Underground water protected

Regulations are promulgated pursuant to the Wyoming statutes Section 35-11-101 to 1104, which states:
"All waters, including ground waters of the State, within the boundaries of the State of Wyoming are the property of the State; and control of the beneficial use of waters of the State resides with the Wyoming State Engineer"

3.3. Groundwater pollution control permit

According to regulations dictated pursuant to Wyoming Statutes 35-11-101 to 1104 (especially 302), stating that a general discharge permit shall be issued prior to discharges of commercial, municipal wastes, and industrial wastes, identified as:

- Municipal wastes
- Chemical, manufacturing and refining wastes
- Mining and mineral processing wastes
- Water produced with oil and gas
- Power generation wastes
- Geothermal fluid or resources wastes
- Nuclear and radioactive wastes
- Toxic and hazardous wastes

Application requirements

A complete application for underground manage for commercial, municipal and industrial wastes shall include:
- Company identification
- Description of the source (chemical, physical, type, radiological and toxic characteristics of discharge), and the receiver (name, geology, hydrology, fluid chemistry, depth of the receiver), and description of testing methods
- Water quality information
- Status of all wells in the area (location and current conditions)
- Information indicating the monitoring and control plans so that discharges will not migrate into the groundwater
- Topography maps
- Volume if the discharge
- Contingency plans

Applications are reviewed by the administrator to determine if it should be accepted as complete or if it requires a notice for public participation; if the application is uncomplete or if it should be denied.

References

Alliance for the Wild Rockies (2017) (https://allianceforthewildrockies.org/) [Verified July, 2017]

ANA - Agência Nacional de Águas (Brazil) (2013). *Water resource management and the mining industry* / National Water Agency, Brazilian Mining Association ; Antônio Félix Domingues, Patrícia Helena Gambogi Boson, Suzana Allpaz, organizers. -- Brasilia: ANA: IBRAM. 334 p.: iii.

Arizona Department of Environmental Quality (2000), ADEQ. Groundwater Protection in Arizona: An Assessment of Groundwater Quality and the Effectiveness of Groundwater Programs

Arizona Department of Environmental Quality (2017). *Registration and Permits*. (http://azdeq.gov/node/529) [Verified July, 2017]

Bridgwood E.W., Singh R.N. and Atkins A.S. (1983) *Selection and optimization of mine pumping system: 1983*. International Journal of Mine Water, Vol. 2 (2). International Mine Water Association.

BCS (2002) *Mining Industry of the Future: Energy and Environmental Profile of the US Mining Industry* (Washington, DC: BCS, Inc.)

Castro JM, Wielinga BW, Gannon JE, Moore JN (1999) *Stimulation of sulfate-reducing bacteria in lake water from a former open-pit mine through addition of organic wastes*. Water Environ. Res. 71, 218-223.

Chilean Copper Commission – COCHILCO (2008). *Best practices and efficient us of water in the mining industry*.

Clark Fork Coalition (2017) (https://clarkfork.org/) [Verified July, 2017]

Colorado Department of Public Health & Environment. Water Quality Control Division. (https://www.colorado.gov/cdphe/wqcd) [Verified July, 2017]

David H. Davis (2014) *Comparing environmental policies in 16 countries /.* - Boca Raton, Fla.: CRC Press

Debie, R.A. (2014). *Comparative Perspectives on Environnemental Policies and Issues*. London: Routledge

Ducks Unlimited (2017) (http://www.ducks.org/) [Verified July, 2017]

Earth First (2017) (http://www.earthfirst.org/) [Verified July, 2017]

Earthwatch Institute (2017) (http://earthwatch.org/) [Verified July, 2017]

Eggert, R. (1994) *Mining and the Environment: International Perspectives on Public Policy*, Washington D.C.: Resources for the Future.

Environmental Defense Fund (2017) (https://www.edf.org/) [Verified July, 2017]

Environmental Science and Engineering / Environmental Science Springer, Berlin, Heidelberg, p. 1 – 10

European Commission, 2007b. Directorate General for Economic and Financial Affairs. European Economy. Economic papers. N298. *Imposing a unilateral carbon constrain on energy intensive industries and its impact on their international competitiveness – Data and analysis*, ed. Office for official publications of the European Communities, Luxemburg.

Ford, L.A., and R.J. McCormick. 2015. Bureau of Land Management. *Water Resource Program Strategy: Focus on Integration, 2015- 2020*. U.S. Department of the Interior, Bureau of Land Management, Washington Office, Washington, DC.

Friends of the Earth (2017) (http://www.foe.org/) [Verified July, 2017]

Furniss, M. J., et al. (2010), *Water, Climate Change, and Forests: Watershed Stewardship for a Changing Climate*, Gen. Tech. Rep. PNW-GTR-812, 75 pp., U.S. Dep. of Agric. For. Serv., Pac. Northwest Res. Stn., Portland, Ore

Gammons, C. H., Harris, L.N., Castro J.M., Cott, P.A., and Hanna, B.W. (2009). *Creating lakes from open pit mines: processes and considerations - with emphasis on northern environments*. Can. Tech. Rep. Fish. Aquat. Sci. 2826: ix + 106 p.

Geller, W., Schultze, M., Kleinmann, R., Wolkersdorfer, C., (2013) (eds.) *Acidic pit lakes - The legacy of coal and metal surface mines*

Greenpeace USA (2017) (http://www.greenpeace.org/usa/) [Verified July, 2017]

Gusek, James J, and Linda A. Figueroa (2009) *Mitigation of Metal Mining Influenced Water*. Littleton, Colo: Society for Mining, Metallurgy, and Exploration.

Hearn, R. AND R. Hoye. (1998) *Copper Dump Leaching and Management Practices That Minimize the Potential for Environmental Releases*. U.S. Environmental Protection Agency, Washington, D.C., EPA/600/2-88/005 (NTIS PB88155114)

Idrysy, Houcyne El and Connelly, Richard (2011). *Water - the other resource a mine needs to estimate*. The 1st International Symposium on Innovation and Technology in the Phosphate Industry. Marrakech, Morocco, May 9- 13, 2011

Janicke M, Weidner H (Eds) (1997) *National Environmental Policies: A Comparative Study of Capacity Building* (Springer, Berlin)

Johnson M S, Cooke J A and Stevenson J K W (1994). *Revegetation of metalliferous wastes and land after metal mining*. In: Hester and Harrison R M (eds.), Mining and Its Environmental Impact.

Kochtcheeva, Lada V. (2009). *Comparative Environmental Regulation in the United States and Russia: Institutions, Flexible Instruments, and Governance*. Albany, NY: SUNY Press

Kochtcheeva, Lada V. (2009) *"Administrative Discretion and Environmental Regulation: Agency Substantive Rules and Court Decisions in U.S. Air and Water Quality Policies."* Review of Policy Research 26 (3).

Lee F.T. and Abel J.F. (1983) *Subsidence from Underground Mining: Environmental Analysis and Planning Consideration*, USGS Circular 876.

Marcus, Jerrold J. (ed) (1997*) Mining Environmental Handbook: Effects of Mining on the Environment and American Environmental Controls on Mining.* Imperial College Press

MonTec (2007) *Guidelines on Financial Guarantees and Inspections for Mining Waste Facilities.* European Commission, DG Environment.

Martin, E.W. (1993) *Environmental Economics & the Mining Industry*, Kluwer Academic Publishers: Amsterdam, W.E. Martin, ed.

Maupin, M.A., Kenny, J.F., Hutson, S.S., Lovelace, J.K., Barber, N.L., and Linsey, K.S., (2014). *Estimated use of water in the United States in 2010*: U.S. Geological Survey

McLemore, Virginia T. (2008). *Basics of Metal Mining Influenced Water* (Management Technologies for Metal Mining Influenced Water, Volume 1). SME.

Mining Magazine, article *"Deepest mine shaft in the US complete"* (May 2016)

Morgan, Stephanie E. (2008). Godwin Pumps. *Dewatering Mines*. Pumps & Systems, September

National Audubon Society (2017) (http://www.audubon.org/) [Verified July, 2017]

National Parks and Conservation Association (2017) (https://www.npca.org/#sm.00012l69j9hgqcotx6k1fk1a1eika) [Verified July, 2017]

National Wildlife Federation (2017) (http://www.nwf.org/) [Verified July, 2017]

Natural Resources Defense Council (2017) (https://www.nrdc.org/) [Verified July, 2017]

Nevada Division of Environmental Protection. Water Pollution control (2016) (https://ndep.nv.gov/water/water-pollution-control) [Verified July, 2017)

Pew Research Center. Views about environmental regulation among adults in Wyoming (2015) (http://www.pewforum.org/religious-landscape-study/state/wyoming/views-about-environmental-regulation/) [Verified July, 2017]

River Network (2017). (https://www.rivernetwork.org/) [Verified July, 2017)

Save Lake Superior Association (2017) (http://www.savelakesuperior.org/) [Verified July, 2017]

Southwest Research and Information Center (2017) (http://www.sric.org/) [Verified July, 2017]

Sea Shepherd Society (2017) (http://www.seashepherd.org/) [Verified July, 2017]

Sierra Club (2017) (http://www.sierraclub.org/) [Verified July, 2017]

Sinclair, J. (1969) *Quarrying, Opencast and Alluvial Mining*, Elsevier, Amsterdam

Sterrett, R. J. (2007) *Groundwater and Wells*, Johnson Division, St.Paul, MN

Sullivan, W and Christian Teo of Purwono & Partners (in association with Stephenson Harwood LLP) (2013) *Mining Law & Regulatory Practice in Indonesia. A primary reference source*. John Wiley & Sons Singapore

The Nature Conservancy (2017) (https://www.nature.org/) [Verified July, 2017]

The Wilderness Society (2017) (http://wilderness.org/) [Verified July, 2017]

Tosun, J. 2013. *Environmental Policy Change in Emerging Market Democracies – Central and Eastern Europe and Latin America Compared*. Toronto: University of Toronto Press

U.S. Army Corps of Engineers (2017). (http://www.usace.army.mil) [Verified, July 2017]

U.S. Bureau of Reclamation (2016). Reclamation Managing Water in the West. (https://www.usbr.gov/) [Verified July, 2017]

U.S. Department of Agriculture (2017). U.S. Forest Service (https://www.fs.fed.us/about-agency/what-we-believe) [Verified July, 2017]

U.S. Department of Energy (2007), *"Mining Industry Energy Bandwidth Study"*.

U.S. Department of Homeland Security (2017). Federal Emergency Management Agency (https://www.fema.gov) [Verified July, 2017]

U.S. Environmental Protection Agency (EPA). 2004. FACTOIDS: Drinking Water and Ground Water Statistics for 2003. EPA-816-K-03-001. Washington, D.C.: Office of Ground Water and Drinking Water, U.S. Environmental Protection Agency

U.S. Fish & Wildlife Service (2013). FWS Fundamentals. (https://www.fws.gov/info/pocketguide/fundamentals.html) [Verified July, 2017]

YOUNGER, P.L. & ROBINS, N.S. (eds) (2002). *Mine Water Hydrogeology and Geochemistry*. Geological Society, London, Special Publications, 198.

World Wildlife Fund (2017) (https://www.worldwildlife.org/) [Verified July, 2017]

Wyoming Department of Environmental Quality. Water Quality. (https://rules.wyo.gov/Search.aspx?mode=1) [Verified July, 2017]

Wyoming State Geological Survey (2016). *Coal Production & Mining.* (http://www.wsgs.wyo.gov/energy/coal-production-mining) [Verified July, 2017]

Zhang, Z., Gao, L., Barrett, D., Chen, Y. *Evaluating water management practice for sustainable mining.* Water. v6 i2. 414-433.

www.ingramcontent.com/pod-product-compliance
Lightning Source LLC
Chambersburg PA
CBHW040217220526
45473CB00001B/25